Varun Gupta

Strategic Value Proposition Innovation Management in Software Startups for Sustained Competitive Advantage

A Strategic Tool for Competitive Advantage

Varun Gupta
Escuela Técnica Superior de Ingenieros
Industriales
Universidad Politecnica de Madrid
Madrid, Spain

Multidisciplinary Research Centre for
Innovations in SMEs (MrciS)
GISMA University of Applied Sciences
Potsdam, Germany

ISSN 1933-978X   ISSN 1933-9798   (electronic)
Synthesis Lectures on Technology, Management, & Entrepreneurship
ISBN 978-3-031-18324-9   ISBN 978-3-031-18322-5   (eBook)
https://doi.org/10.1007/978-3-031-18322-5

This Springer imprint is published by the registered company Springer Nature Switzerland AG
The registered company address is: Gewerbestrasse 11, 6330 Cham, Switzerland

My research is dedicated to my parents as well as the almighty Goddesses Koyla, Myah Bhagwati, Jalpa Mata, and Naina Mata.

Mi investigación está dedicada a mis padres y a las diosas todopoderosas Koyla, Myah Bhagwati, Jalpa Mata y Naina Mata.

Meine Forschung ist meinen Eltern sowie den allmächtigen Göttinnen Koyla, Myah Bhagwati, Jalpa Mata und Naina Mata gewidmet.

Preface

The business environment is highly volatile and competitive. The issue had been exacerbated by the abrupt breakout of a pandemic, which had caused global economic chaos and significant oscillations in the business environment. Entrepreneurship is perilous, hard, and challenging in ever-changing business conditions, particularly in the software industry.

In highly competitive marketplaces, a startup's success is dependent on the entrepreneur's continuous business model innovation initiatives. Startup capacity to foster business model improvements is hampered by their liability of newness and smallness. With these startups, open innovation becomes a viable option for fostering innovation. Value proposition innovation is the most important factor that leads to business model innovation, which leads to higher performance levels and competitive advantage in a highly competitive environment. If startups can strengthen their dynamic capacities to drive value proposition innovation, they may be able to maintain a competitive advantage for extended periods of time. Exploratory research is required due to the scarcity of research in the value proposition innovation domain in the setting of software startups.

Growing trends in the gig economy, technological innovations in freelancing platforms and online feedback acquisition tools, increased focus on customer-centric innovations driven by co-creation with customers, and increased entrepreneurial orientation in academia are all factors that point to opportunities for startups.

According to the dynamic capabilities theory, startups should use their dynamic capabilities to continuously renew their ordinary and operational capabilities as well as resources in order to achieve superior performance over long periods of time in a highly fluctuating software industry environment. In the context of a startup, the capacity to enhance dynamic capabilities is mostly based on organizational flexibility, as they have limited strategic resources. There is a need to investigate whether open innovation elements such as freelancers, customers, and academia could be integrated into startups as strategic resources for value proposition innovation, resulting in synergism with startup organizational flexibility for strengthening dynamic capabilities.

The research investigates the problem of *how the software startups could foster value proposition innovation by leveraging open innovation with customers, freelancers, and academia for sustained competitive advantage.*

Multiple research approaches were included in the study plan and conducted to solve the research problem and overcome the minimal support from the body of literature about value proposition innovation in the setting of software startups. Systematic mapping studies, multiple case studies, surveys, experience reports, and their combinations were among the several research approaches used. Systematic mapping studies aided in the analysis of research trends in the value proposition innovation domain, as well as freelancing participation in such innovations in the startup environment. The findings of mapping studies aided in the conceptualization of frameworks for other research studies aimed at examining how freelancers, customers, and academia may be involved in value proposition innovation. By exploring the phenomena in a real-world setting, several case studies investigated freelancer-driven value proposition innovation in companies. Using online feedback-gathering methods, experience reports were used to investigate how customers may be involved in value proposition innovations using online feedback acquisition technologies. Furthermore, throughout the epidemic, practical experiences explored how the collaboration of freelancers, academia, and customers may assist startups in exploring overseas markets with minimal resources. In addition, the results of the research approach give the necessary theoretical framework for the next step. The research outcomes were validated through member checking using surveys to improve the studies' reliability. Small surveys were used in certain studies to quantitatively measure the phenomenon under examination. Together, qualitative, and quantitative studies add more realism to the research and make them more exploratory.

According to the findings, startups can reinvent value propositions by integrating freelancers for both domestic and global markets. They might use freelancers employing panel-based, task-based, and hybrid ways that could be designed in a crowdsourced or non-crowdsourced way, depending on their internal competencies and external environment analyses, such as threats and opportunities. However, strategic partnerships with freelancers must be structured in such a way that both parties profit. Lower development costs, decreased time to market, reduced time to product/market fit, and greater customer perceived value have all been claimed as benefits of these associations. Rather than relying entirely on freelancers for outsourced services, startup teams and freelancers should collaborate on multiple value proposition activities as a team to ensure two-way knowledge sharing. Furthermore, freelancers' involvement could aid startups in gaining customer perspectives based on their understanding of customers as a result of their close proximity to them. Startups could potentially gain access to global talent by strategically utilizing freelancing platform technologies. These platforms can be useful for finding worldwide talent, gaining access to foreign market intelligence, gaining access to global customer segments, and expanding a startup's panel of freelancers. Access to these platforms should be a part of the overall strategy for digitally integrating talent into the startup team.

In order to be successful, customers must be involved in the innovation process. Startups can overcome the liability of newness and smallness in gaining access to global clients in a variety of ways. First, the startup might focus down to significant client segments with the help of academia, experts, and freelancers. This will allow them to concentrate their efforts on early adopters solely. Experts' reputations and proximity to clients could aid startups in bringing their professional proximities closer to customers. Second, by incorporating online feedback acquisition technology into startup business activities such as social networking sites, customers' awareness, interests, desires, and actions toward the product can be boosted. These technologies assist startups in bringing their professional proximities closer with the customers. These businesses may make the most of freelancing and online feedback-gathering technology. This enables startups to gain access to international markets and expand their business operations. Furthermore, academia might be an excellent partner in developing new value propositions. This, however, necessitates the creation of a shared-values environment that aids each partner in achieving their business objectives.

If startups strategically partner with freelancers, customers, and academia, they can assist startups drive value proposition innovation and transform them into strategic resources. This necessitates startups focusing on strategic partnerships as part of their overall business strategy rather than forming partnerships solely to meet short-term probabilistic outsourcing demands. In order to sustain a competitive advantage, startups must cultivate, maintain, and enhance strategic relationships with freelancers, customers, and academia in order to respond to the always changing software industry environment by continually building dynamic capabilities. Their ongoing, mutually beneficial participation in innovation activities aids startups in transforming them into strategic resources and obtaining their assistance for knowledge acquisition and exploitation in order to develop their dynamic capabilities. Value proposition innovation is a competitive advantage tool, but the long-term competitive advantage is contingent on continual innovation. Startups must overcome their liabilities of newness and smallness in order to consistently reinvent value propositions, which necessitates strategic open innovation with freelancers, customers, and academia. This increases the startups' dynamic capabilities in responding to changing business environments through value proposition innovation and achieving a sustainable competitive edge. Figure 1 depicts a comprehensive view of the book's main theme.

In two ways, this argument advances the theory: first, by expanding the corpus of knowledge, and second, by introducing new approaches to encouraging value proposition creation in a highly dynamic context that differ from what has previously been done. The book adds to the body of knowledge in the fields of management and technical sciences about value proposition innovation. Despite the fact that the findings are mainly management oriented, technical researchers will find them valuable in bridging the gaps through technical improvements, such as computationally enhanced algorithms or tools. As a result, managers, entrepreneurs, and technological researchers will be able to share information in a two-way fashion.

This book also included methodological contributions. The methodological innovation stems from the application of numerous research methods in real-world contexts, the use

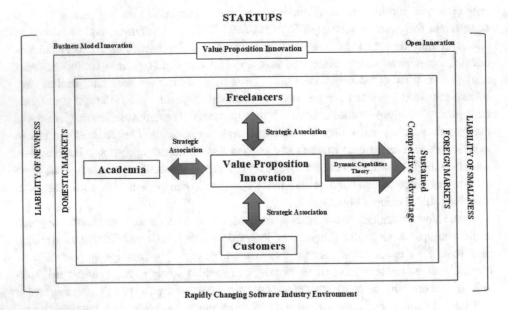

Fig. 1 Holistic view of book focus

of both qualitative and quantitative data, and the careful selection of good startup samples. Furthermore, the concept has ramifications for business owners, customers, freelancers, and academics. Entrepreneurs should use the findings to improve their business methods, such as using online feedback-gathering technology, freelancing platform technology, and freelancer associations. Customers, freelancers, and academics can use the results to assess a company and make informed judgments about their long-term relationship with them.

In the research fields of strategy management, entrepreneurship, globalization, business model innovation, and innovation management, the book outputs are more management oriented. They do, however, have interdisciplinary applications, such as in software engineering, natural language processing, and artificial intelligence.

Keywords Software startups · Entrepreneurship · Open innovation · Business model innovation · Value proposition innovation · Freelancers · Gig economy · Freelancing platform technology · Customers · Online feedback acquisition technology · Academia · Strategic associations · Sustained competitive advantage · Dynamic capabilities theory

Madrid, Spain   Varun Gupta
Potsdam, Germany

Contents

List of Figures

List of Tables

Introduction

1

This introductory chapter of the book aims to advance the understanding of value proposition innovation in software startups driven by strategic support of innovation ecosystem
elements involving freelancers and academia. This chapter first introduces the research
background and the motivation for conducting the research as disseminated in this book.
The research background empirically establishes the research context and well positions
the research problem. After that research objectives and research questions are presented.
They present the research outcomes that will be achieved after the research finds answers
to formulated research questions. After that, the structure of the book is presented. The
graphical model is then presented that represents the research problem addressed. Finally,
the contribution of the book in the creation of new knowledge in value proposition innovation in resource-constrained startups is presented. This new knowledge is useful to
entrepreneurs, customers, and the nation's economy.

1.1 Research Background and Motivation

The startups have an impact on the nation's economy (Praag & Versloot, 2007; Jo &
Jang, 2019). They provide innovative products in the market that foster innovations, create employment, improve the productivity of other industries through their innovative
offerings and create opportunities for foreign trade. They lead to economic growth which
is reflected in improvements in measures like gross domestic product (GDP), inflation
rates, employment rates, production facilities, regional, developments, and much more
(Gupta & Rubalcaba, 2021). The economic contribution of startups had been well recognised by federal governments as well. For instance, the European Union (EU) had been
taking larger initiatives to boost innovation and entrepreneurship, for instance, Erasmus

© The Author(s), under exclusive license to Springer Nature Switzerland AG 2022
V. Gupta, *Strategic Value Proposition Innovation Management in Software Startups
for Sustained Competitive Advantage*, Synthesis Lectures on Technology, Management,
& Entrepreneurship, https://doi.org/10.1007/978-3-031-18322-5_1

for Young Entrepreneurs programme (European Commission, 2021a) and making the process of starting a new business easier (European Commission, 2021b).

A startup is an organization founded to find a repeatable and scalable business model (Blank, 2013, 2020). To put it another way, startups are transient businesses that are always trying to find a scalable and repeatable business model. Once the scalable and repeatable business model is identified, the level of experimentation lowers leading to fewer changes in the corresponding business model. However, to sustain a competitive advantage, the startup should continuously innovate their business models. This could take the form of value proposition innovation where the value proposition canvas is continuously evolved to enhance customer value. For instance, the product could be evolved by increasing its performance, enhancing its functionality, or adding more security features, which adds to customer value.

Business model innovation leads to competitive advantage and higher performance levels (Eggers & Kraus, 2011; Vargo & Seville, 2011; Kraus et al., 2020a, 2020b; Albats et al., 2021; Clauss et al., 2021; Latifi et al., 2021). Value proposition innovation assists startups in identifying market opportunities (Guo et al., 2021), creating new markets or expanding existing ones (Guo et al., 2021), increasing customer retention rates (Chesbrough, 2007), improving customer relationships (Guo et al., 2021), and exploring entrepreneurial opportunities in highly uncertain environments (Guo et al., 2021). (Chesbrough, 2010; Clauss, 2017; Guo et al., 2021). Improved firm performance (Chandler et al., 2014; Covin et al., 2015), competitive advantage (Antonopoulou & Begkos, 2020; Rintamäki & Saarijärvi, 2021), and increased revenues are all business benefits of value proposition innovation (Chesbrough, 2007).

The value proposition innovation is an external dynamic capability of startups that helps them respond to the continuously changing business environment (Guo et al., 2021; Schmidt & Scaringella, 2020; Teece et al., 1997). Further, value proposition innovation could result in disruptive innovations (Schmidt & Scaringella, 2020). However, the liability of newness and the liability of smallness limit their potential to develop such advances (Gama et al., 2019; Gimenez-Fernandez et al., 2020). The reason is that innovations require startups to explore the opportunities and trends across the business environment, including the market, to select the most valuable ideas, implement and finally commercialise them. Innovation also requires firms to be highly flexible as well as responsive to environmental opportunities. This means that startups can stimulate innovation by leveraging their agility, flat organizational structures, and quick decision-making abilities (Nooteboom, 1994; Rosenbusch et al., 2011; Vossen, 1998).

The startups could be disruptive innovators in the market if they take the best possible advantage of these positive aspects that arise because of their organizational setups by optimally making use of their resources as well as strategically extending their resources to overcome the liability of newness and smallness.

Software startups focus on developing innovative software intensive products or services (Giardino et al., 2014; Paternoster et al., 2014; Rafiq et al., 2021). These innovative solutions help customers to adopt these technologies to solve their unmet needs. For example, Uber, Facebook, and Netflix had been disruptive software innovations that offer services to customers. Innovation is all about commercialising creative ideas; ideas that are new and valuable; solving customers' unmet problems and providing them with expected benefits.

Software startups share all the challenges as introduced before. However, software startups face more challenges compared with other startups; startups involved in other non-software intensive products or services. This includes-higher uncertainty levels because of technological fluctuations in the software industry (Unterkalmsteiner et al., 2016) and the adoption of software development process models of matured companies in their context (Giardino et al., 2014; Unterkalmsteiner et al., 2016; Klotins et al., 2019; Saad et al., 2021), which signify that implementing innovative ideas for commercialisation is harder. Limited resources, especially experience and human resources, make the implementation harder. Technological fluctuations could result in competitors coming up with better technological solutions or cost reductions driven by the adoption of innovative technology. These technological fluctuations could result in difficulties to retain a competitive advantage or even complete failure in the market. In particular, the innovation challenges are much amplified in the context of startup context for inventing new ideas, implementation, and commercialisation.

Fostering value proposition innovation driven by openly innovating with freelancers, customers and academia could assist startups to overcome their liabilities of newness and smallness in order to stimulate innovation (Hilmersson & Hilmersson, 2021; Kraus et al., 2020a, 2020b; Rosenbusch et al., 2011). Gig economy is growing and had resulted in continuous innovations and an increase in the number of freelancing platform technologies (Fuller et al., 2020). Freelancers could help businesses to find new capabilities, increase labour force flexibility, increase speed to market, and innovate new business models (Adamson, 2021). Involving freelancers optimally across platforms and directly seems to be an interesting research domain for fostering value proposition innovations in software startups. The involvement of freelancers across different value proposition innovation activities is another interesting avenue for research.

The identification of product value proposition through customer involvement is normally conducted face to face in co-located physical space. This is grounded in the customer development theory (Blank, 2013, 2020), which focuses on "going out of building and talking with customers" and is reported to be the reason for meeting customers actual needs (Newbert et al., 2020). This type of interaction is an effortful activity and cannot be applied if the startup is looking for internalization. These effortful interactions could draw a thin line between business continuity and burnout for the startups. This interaction is infeasible as well during pandemics due to imposed lockdowns and other social distancing norms in place, forcing the entrepreneurs to adopt other means of interactions like adopting online feedback acquisition tools. However, this adoption during the pandemic time was not an easy task as the support from the scholarly literature was too limited. Further, deploying computationally advanced infrastructures, for instance, crowd-sourcing based ones (Lim & Finkelstein, 2012; Renzel et al., 2013; Snijders et al., 2015; Groen et al., 2017) for customer feedback seems not feasible in the context of resource stripped startups. The ability to identify the compelling value proposition through an active customer participation becomes more challenging with limited resources that startups own and the market innovativeness i.e., the higher the innovativeness, the lower is market understanding and more is the need for customer interaction.

Product value proposition is a tool to sustain competitive advantage; the task for which the involvement of customers is of utmost importance throughout the product lifecycle to get continuous product-related feedback. The challenge to explore markets through co-located and remote interactions with customers with limited resources need innovative ways of gathering their perspectives that are not only accurate but cost-efficient as well. The role of customers in fostering innovation had already been reported in (Desouza et al., 2008; Lindič & Silva, 2011; Melander, 2019; Taghizadeh et al., 2018; Wijekoon et al., 2021). This is true as the success or failure of startups is determined by the repeated purchases made by the customers. Whether the product is innovative or not, is determined by their adoption by the customers. Involving customers in value proposition innovation especially during the pandemic is another interesting research area to be explored by this book.

The strategic partnerships especially with Academia could provide startups with the necessary resources for exploring customer markets, taking advantage from university support for businesses. The experts, students and libraries could provide meaningful perspectives about the expected customer needs and provide resources to implement and commercialise the product offering to the market. The academia support could be a long term and sustained one, helping startups to continuously evolve their product as per learning obtained from the markets. The academic resources, especially experts and students, could be integrated as freelancers for continuing startup operations.

The co-innovation with freelancers, customers, and academia for fostering innovation is in its infancy stage and needs to be empirically explored to provide real value to the startup community. The research will involve systematic reviews of literature, case studies, surveys, and experience reports to uncover the freelancer and customer involvement in value proposition innovation taking best possible advantage of available technologies. The adoption of freely available technologies could help startups to foster value proposition innovation, but it needs further research to investigate the factors leading to such adoption so that the technologies satisfying these factors be adopted for startups.

Business model innovation literature in startup context is still an underdeveloped issue (Ibarra et al., 2020; Pucihar et al., 2019). Value proposition is also an underdeveloped concept in literature (Antonopoulou & Begkos, 2020) with limited research conducted in this domain (Schmidt & Scaringella, 2020).

The overall impact of the freelancer, customer and academic involvement in value proposition innovation is to attain sustained competitive advantage. The startup sustained competitive advantage can be well explained using dynamic capabilities theory (Teece et al., 1997), which is extended from the resource based view as proposed in (Barney, 1991). This is because, this theory explains that how firms could have long term competitive advantage based on their dynamic capabilities in highly uncertain environments. As startups operate in continuously changing software industries and they need to innovate fast, the dynamic capabilities theory is a suitable option.

The business model innovation is triggered by dynamic capabilities of the firm (Čirjevskis, 2019; Teece, 2018). This is because, the process of innovating business models requires dynamic capabilities to sense opportunities, seize the opportunities and transform. Value proposition innovation is also a way to make a response to the fluctuating business environment using the resource base of the startup and hence is external dynamic capability of the startup (Teece et al., 1997). Value proposition innovation, which is central element of business model innovation, requires startups to own dynamic capabilities to identify, implement and commercialise it (Schmidt & Scaringella, 2020). Further, value proposition innovation could be considered as an external dynamic capability which explores market opportunities, seize the opportunities by mobilising resources to implement ideas followed by continuous renewal (Teece, 2018). The flexibility in startup organizational setup positively impacts its dynamic capabilities but limited scare resources limit these capabilities (Fabrizio et al., 2021).

Strategic innovation, according to Markides (1997), is what allows new businesses to compete successfully in markets with established players. This is due to the fact that strategic innovation is what allows new businesses to "break" the rules of the game, i.e., prevalent business trends, and develop entirely novel methods of conducting business and delivering unique goods or services. For instance, these companies might devise brand-new strategies for product distribution that are entirely different from those used by the market's current leaders. Another example is that these companies might commercialize a brand-new method of making a product that is not yet used by other participants in the market. Alternately, the new businesses can develop a totally unique value proposition from what their rivals are now providing. These few examples demonstrate how established companies could violate business norms rather than adopt traditional strategies to unseat market leaders. By challenging the established standards, new businesses are offering clients creative value propositions that boost value and reduce expenses. Thus, strategic innovation is a factor in the successful market entry of new companies into markets that would have otherwise been dominated by their rivals. On the same line, this book defines strategic value proposition innovation management as innovative methods for implementing value proposition innovation that is fully distinct from those used by its rivals. By challenging current methods of doing this innovation management, the strategic value proposition innovation management approach aids in the identification, implementation, and commercialization of innovative value propositions. The strategist's main goal is to carry out innovation in a methodical way with a long-term perspective. The strategic value proposition innovation management method, which is highlighted in this book, involves customers, academics, and freelancers in locating, putting into practice, and commercializing the innovative value proposition. The companies have a long-term goal in mind to successfully innovate in markets, and they have strategic ties with them as well as ongoing experiments to raise these partnerships to greater heights. The new approaches to managing value proposition innovation, which involve customers, academics, and freelancers, provide new rules to the open innovation game and are strategic in nature. The term "value proposition innovation management" is used throughout the rest of the book to refer to strategic value proposition innovation management.

The limited research in business model innovation and value proposition innovation; higher startup failure rates and the importance of these innovations in sustained startup market success provided necessary gaps that need to be explored in this book. Further, growing innovations and positive trends across the gig economy, co-creation with customers, and academia support for entrepreneurship provide research directions to bridge the identified research gaps. The value proposition innovation thus needs support from an innovation ecosystem that is sustainable and valuable for the startups. From the lens of dynamic capabilities theory, it is worthy to investigate if freelancer, customer, and academia participation could aid a startup in building its dynamic capabilities to foster value proposition innovation management for long term competitive advantage.

1.2   Research Objectives and Research Questions

The overall research objective (RO) is to increase the success rates of startups by focusing on value proposition innovation, which is propelled by the involvement of potential consumers as well as other resources such as freelancers and strategic relationships with academia. This research objective can be decomposed into four sub-objectives (RO1, RO2, RO3 and RO4) as formulated below:

RO1. To discover the current state of value proposition innovation and the need for additional research efforts.

RO2. To discover the current state of freelancer driven value proposition innovation and the need for additional research efforts.

RO3. To empirically explore and strategically build the associations of freelancers with startups in real practice for value proposition innovation activities.

RO4. To empirically explore the co-creation with customers and strategic partnerships with academia to foster value proposition identification in startups.

To achieve these objectives, the book tries to find answers to the research questions about co-innovation supported by freelancers, customers, and strategic partnerships in a startup context. The research questions are formulated as under:

RQ1. What is the current state of value proposition innovation in startups?

RQ2. What is the current state of freelancer participation in startup value proposition innovation management?

RQ3. What is the current state of practice in startups for freelancer participation in innovation management activities in terms of methodologies, practices, problems, and real-world outcomes?

RQ4. How can value proposition innovation be fostered through customer participation?

RQ5. How can entrepreneurs assess freelancing platforms to create and sustain strategic relationships with freelancers?

RQ6. How might academic strategic partnerships be investigated as a means of promoting innovation management?

To achieve the research objective, this book decomposes the research study into six independent but related research studies. Each research study corresponds to individual research questions as formulated in this chapter. For instance, research study 1 aims to investigate the current state of value proposition innovation in startups. Research study 2 aims to investigate the state of freelancer participation in startup innovation management activities. The focus here is to investigate practices based on synthesis of the bibliographic literature. Research study 3 aims to investigate the current state of freelancer participation in startup innovation activities in terms of methodologies, practices, problems, and real-world outcomes. The focus here is on investigating the practices in startup context. Research study 4 aims to investigate how customer participation could foster value proposition innovation. Research study 5 aims to investigate how entrepreneurs can assess freelancing platforms to create and sustain strategic relationships with freelancers. Research study 6 aims to explore how academic strategic partnerships can be investigated as a means of promoting innovation management. Table 1.1 represents the research questions addressed by individual research studies.

Figure 1.1 shows the relationship between research objectives and research questions.

1.3   Structure of Book

This book is composed of six chapters which represent different research articles as highlighted in Sect. 1.5. The articles are the outcomes of the research conducted to meet research objective by answering the research questions as formulated in the book.

This chapter introduces the research area, the research gaps, research objectives, and research questions that will be targeted by the invested research efforts. This chapter focuses on making readers aware of the research background and the problems that will be targeted by the research. This chapter introduces the contextual information leading to better understanding of the research conducted and disseminated in the book.

Table 1.1 Research studies and research objectives

S. No.	Research study	Research objective	Research question
1	Study 1	RO1. To discover the current state of value proposition innovation and the need for additional research efforts	RQ1. What is the current state of value proposition innovation in startups?
2	Study 2	RO2. To discover the current state of freelancer driven value proposition innovation and the need for additional research efforts	RQ2. What is the current state of freelancer participation in startup innovation management activities?
3	Study 3	RO3. To empirically explore and strategically build the associations of freelancers with startups in real practice for value proposition innovation activities	RQ3. What is the current state of practice in startups for freelancer participation in innovation management activities in terms of methodologies, practices, problems, and real-world outcomes?
4	Study 4	RO4. To empirically explore the co-creation with customers and strategic partnerships with academia to foster value proposition identification in startups	RQ4. How can value proposition innovation be fostered through customer participation?
5	Study 5	RO3. To empirically explore and strategically build the associations of freelancers with startups in real practice for value proposition innovation activities	RQ5. How can entrepreneurs assess freelancing platforms to create and sustain strategic relationships with freelancers?
6	Study 6	RO4. To empirically explore the co-creation with customers and strategic partnerships with academia to foster value proposition identification in startups	RQ6. How might academic strategic partnerships be investigated as a means of promoting innovation management?

Chapter 2 provides the theoretical background related to importance of business model innovation in general and value proposition innovation in particular for software startups. Driven by literature review, research gaps in the value proposition innovation domain are identified. The review sets the motivation for further research in fostering innovations driven by involving freelancers, customers, and academia.

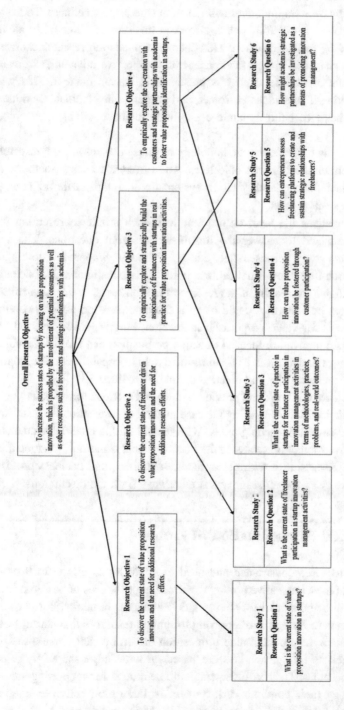

Fig. 1.1 Relationship between research objectives, research studies and research questions

The overall efforts will be to increase startup success rates driven by their ability to foster innovations by overcoming their liabilities of newness and smallness. This is accomplished by empirically investigating existing work in this research domain to identify the gaps to be bridged by further research. This chapter expands the research motivation as introduced in this chapter. Further, the concept of competitive advantage from the perspective of dynamic capabilities theory is presented in startup context. This will form the basis of predicting if freelancer, customer, and academia involvement in value proposition innovation helps to build dynamic capabilities for having long term competitive advantage.

Chapters 3 presents the research methodology as chosen to answer the research questions and meet formulated research objectives. The different research methods that are executed to conduct the research include systematic literature reviews, case studies, surveys, and experience reports.

Chapter 4 summarises the research articles; with each article representing the individual studies conducted to answer the formulated research questions. The summary outlines the research study purpose, study design/methodology/approach, findings, originality/value, research limitations/implications, practical implications, and conclusion.

Chapter 5 highlights the results of the research for each research question followed by the discussion about how startups could make a strategic decision about involving freelancers and academia for value proposition innovation. Further, it is discussed that how such a strategic open innovation focused on value proposition can lead to sustained competitive advantage grounded on strengthening of dynamic capabilities. The contribution of book to theory, methodology and practice is finally provided.

Chapter 6 concludes the book and provides strategic directions for book audience. This book aids academic institutions in using the case method of instruction because Appendix includes a genuine case and teaching notes. With the aid of this material, students will be able to apply the knowledge presented in the book to actual situations, practice their business creativity by discussing it with peers, faculty members, and business professionals, and ultimately take away the most important takeaways from the discussions.

1.4 Graphical Representation of Problem

The startups are resource constrained and need to invest lot of efforts in exploring the potential market of their innovative products with the high chances of burn-out. Exploring global markets will be beneficial for startup quicker success in market but is challenging due to limited access to foreign market owing to limited resources that startups own. The possibility is to innovate product value proposition driven by richer market exploration in domestic as well as foreign markets are based on knowledge shared by freelancers, customers, and academia. Their involvement could be crucial for supporting value proposition innovation by their continues involvement in innovation activities namely ideas generation, implementation, and commercialisation. Their support could help the startups

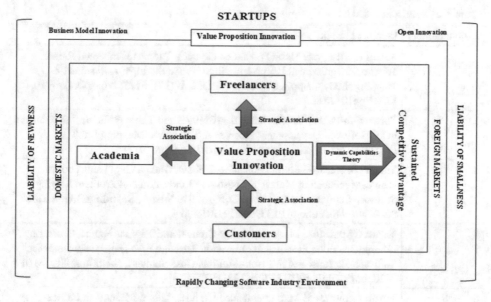

Fig. 1.2 Graphical representation of research problem

to execute open innovation for value proposition based on richer analysis of the business environment overcoming the factors that hinder their innovation potential. Focusing on value proposition innovation will help startup to enhance its firm's performance and gain competitive advantage.

The problem is thus defined as "*How the software startups could foster value proposition innovation by leveraging open innovation with Customers, freelancers and academia for sustained competitive advantage*".

The ability of startups to foster value proposition innovation should result into long term competitive advantage in a highly fluctuating business environment. The startups need dynamic capabilities to have long term superior performance relative to their competitors in industry. Open innovation with freelancers, customers, and academia should help startup strengthen its dynamic capability to be able to make market success in highly fluctuating software industry. Figure 1.2 graphically represents the problem statement targeted by this book.

1.5   Scientific Contributions Forming Part of the Book

This book section lists the different outcomes of the research study as disseminated in various Journals and Conferences of repute. This book is based on following articles which could be explored for further reading. These further reading material together contribute to the research objectives of the doctoral research as disseminated in this book (Table 1.2).

Table 1.2 Publication details

Article number	Details of article
1	**Varun Gupta**, José María Fernández-Crehuet, Thomas Hanne and Rainer Telesko, "Requirement Engineering in Software Startups: A Systematic Mapping Study", **Applied Sciences**, MDPI, 10(17), 6125; https://doi.org/10.3390/app10176125
2	**Varun Gupta**, José María Fernández-Crehuet and Thomas Hanne, "Freelancers in the Software Development Process: A Systematic Mapping Study", Processes, MDPI, 2020, https://doi.org/10.3390/pr8101215
3	**Varun Gupta**, José María Fernández-Crehuet, Chetna Gupta and Thomas Hanne, "Freelancing Models for Fostering Innovation and Problem Solving in Software Startups: An Empirical Comparative Study", **Sustainability**, MDPI, 2020, http://dx.doi.org/10.3390/su122310106
4	**Varun Gupta**, José María Fernández-Crehuet and Thomas Hanne, "Fostering Continuous Value Proposition Innovation Through Freelancer involvement in Software Startups: Insights from Multiple Case Studies", **Sustainability**, MDPI, 2020, https://doi.org/10.3390/su12218922
5	**Varun Gupta**, José María Fernández-Crehuet, Thomas Hanne and Rainer Telesko, "Fostering Product Innovations in Software Startups through Freelancer Supported Requirement Engineering", **Results in Engineering, Elsevier,** 2020, 8, https://doi.org/10.1016/j.rineng.2020.100175
6	**Varun Gupta** and José María Fernández-Crehuet, Online Feedback Management Tools for Early-Stage Startups: Hidden Treasures in Rocky Mountains, **IT Professional**, IEEE, 2021 https://doi.org/10.1109/MITP.2021.3105253
7	**Varun Gupta**, José María Fernández-Crehuet and Chetna Gupta, "Freelancer Supported Requirement Engineering Framework for Software Start-Ups", **IEEE Computer**, 2022, https://doi.org/10.1109/MC.2022.3180711
8	**Varun Gupta** and José María Fernández-Crehuet, "Book Review on "Networks, SMEs, and the university: The process of collaboration and open innovation", **International Small Business Journal: Researching Entrepreneurship**, SAGE, pp: 1–3, 2021, https://doi.org/10.1177/02662426211008009
9	**Varun Gupta**, José María Fernández-Crehuet, "Divergent Creativity for Requirement Elicitation Amid Pandemic: Experience from Real Consulting Project", "Tenth International Workshop on Creativity in Requirements Engineering (CreaRE'21)", **The 27th International Working Conference on Requirement Engineering: Foundation for Software Quality (REFSQ 2021)**, 12th April 2021, Essen, Germany
10	**Varun Gupta**, José María Fernández-Crehuet and Dariusz Milewski, "Academic-Startup Partnerships to Creating Mutual Value", **IEEE Management Review**, IEEE, Volume 49, Issue 2, https://doi.org/10.1109/EMR.2021.3065276

References

Adamson, C. (2021). How freelance developers can help your midsize business. *Harvard Business Review*.

Albats, E., Podmetina, D., & Vanhaverbeke, W. (2021). Open innovation in SMEs: A process view towards business model innovation. *Journal of Small Business Management*. pp. 1–42 https://doi.org/10.1080/00472778.2021.1913595

Antonopoulou, K., & Begkos, C. (2020). Strategizing for digital innovations: Value propositions for transcending market boundaries. *Technological Forecasting and Social Change, 156*, 120042.

Barney, J. B. (1991). Firm resources and sustained competitive advantage. *Journal of Management, 17*(1), 99–120.

Blank, S. (2013). Why the lean start-up changes everything. *Harvard Business Review, 91*(5), 63–72.

Blank, S. (2020). *The four steps to the epiphany: Successful strategies for products that win*. Wiley.

Chandler, G. N., Broberg, J. C., & Allison, T. H. (2014). Customer value propositions in declining industries: Differences between industry representative and high-growth firms. *Strategic Entrepreneurship Journal, 8*(3), 234–253.

Chesbrough, H. (2007). Business model innovation: It's not just about technology anymore. *Strategy & Leadership, 35*(6), 12–17.

Chesbrough, H. (2010). Business model innovation: Opportunities and barriers. *Long Range Planning, 43*(2), 354–363.

Čirjevskis, A. (2019). The role of dynamic capabilities as drivers of business model innovation in mergers and acquisitions of technology-advanced firms. *Journal of Open Innovation: Technology, Market, and Complexity, 5*(1), 12. https://doi.org/10.3390/joitmc5010012

Clauss, T. (2017). Measuring business model innovation: Conceptualization, scale development, and proof of performance. *R&d Management, 47*(3), 385–403.

Clauss, T., Breier, M., Kraus, S., Durst, S., & Mahto, R. V. (2021). Temporary business model innovation–SMEs' innovation response to the Covid-19 crisis. *R&D Management*. https://doi.org/10.1111/radm.12498

Eggers, F., & Kraus, S. (2011). Growing young SMEs in hard economic times: The impact of entrepreneurial and customer orientations—A qualitative study from Silicon Valley. *Journal of Small Business & Entrepreneurship, 24*, 99–111.

European Commission. (2021a). *Erasmus for young entrepreneurs*. https://ec.europa.eu/growth/smes/supporting-entrepreneurship/erasmus-young-entrepreneurs_en

European Commission. (2021b). *Start-up procedures*. Available online: https://ec.europa.eu/growth/smes/sme-strategy/start-up-procedures_en.

Covin, J. G., Garrett, R. P., Jr., Kuratko, D. F., & Shepherd, D. A. (2015). Value proposition evolution and the performance of internal corporate ventures. *Journal of Business Venturing, 30*(5), 749–774.

Desouza, K. C., Awazu, Y., Jha, S., Dombrowski, C., Papagari, S., Baloh, P., & Kim, J. Y. (2008). Customer-driven innovation. *Research-Technology Management, 51*(3), 35–44. https://doi.org/10.1080/08956308.2008.11657503

Fabrizio, C. M., Kaczam, F., de Moura, G. L., da Silva, L. S. C. V., da Silva, W. V., & da Veiga, C. P. (2021). Competitive advantage and dynamic capability in small and medium-sized enterprises: A systematic literature review and future research directions. *Review of Managerial Science*, 1–32 https://doi.org/10.1007/s11846-021-00459-8

Fuller J. B., Raman M., Palano J., Bailey A., Vaduganathan N., Kaufman E., Laverdière, R., & Lovett, S. (2020). *Building the on-demand workforce*. Harvard Business School and BCG.

Gama, F., Frishammar, J., & Parida, V. (2019). Idea generation and open innovation in SMEs: When does market-based collaboration pay off most? *Creativity and Innovation Management, 28*(1), 113–123. https://doi.org/10.1111/caim.12274

Giardino, C., Unterkalmsteiner, M., Paternoster, N., Gorschek, T., & Abrahamsson, P. (2014). What do we know about software development in startups? *IEEE Software, 31*(5), 28–32.

Gimenez-Fernandez, E. M., Sandulli, F. D., & Bogers, M. (2020). Unpacking liabilities of newness and smallness in innovative start-ups: Investigating the differences in innovation performance between new and older small firms. *Research Policy, 49*(10). https://doi.org/10.1016/j.respol.2020.104049

Groen, E. C., Seyff, N., Ali, R., Dalpiaz, F., Doerr, J., Guzman, E., Hosseini, M., Marco, J., Oriol, M., Perini, A., & Stade, M. (2017). The crowd in requirements engineering: The landscape and challenges. *IEEE Software, 34*(2), 44–52. https://doi.org/10.1109/MS.2017.33

Guo, H., Yang, J., & Han, J. (2021). The fit between value proposition innovation and technological innovation in the digital environment: Implications for the performance of startups. *IEEE Transactions on Engineering Management, 68*(3), 797–809. https://doi.org/10.1109/TEM.2019.2918931

Gupta, V., & Rubalcaba, L. (2021). Competency-Industry Relatedness (C-IR) framework for sustained business growth in startups during and beyond pandemic: Myths and lessons from publicly funded innovative startups. Sustainability, 13(9), 4632.

Hilmersson, F. P., & Hilmersson, M. (2021). Networking to accelerate the pace of SME innovations. *Journal of Innovation & Knowledge, 6*(1), 43–49. https://doi.org/10.1016/j.jik.2020.10.001

Ibarra, D., Bigdeli, A. Z., Igartua, J. I., & Ganzarain, J. (2020). Business model innovation in established SMEs: A configurational approach. *Journal of Open Innovation: Technology, Market, and Complexity, 6*(3), 76. https://doi.org/10.3390/joitmc6030076

Jo, G. S., & Jang, P. (2022). Innovation characteristics of high-growth startups: The Korean case startups. *Journal of Small Business & Entrepreneurship, 34*(2), 222–239.

Klotins, E., Unterkalmsteiner, M., & Gorschek, T. (2019). Software engineering in start-up companies: An analysis of 88 experience reports. *Empirical Software Engineering, 24*(1), 68–102 (2019). https://doi.org/10.1007/s10664-018-9620-y

Kraus, S., Clauss, T., Breier, M., Gast, J., Zardini, A., & Tiberius, V. (2020a). The economics of COVID-19: Initial empirical evidence on how family firms in five European countries cope with the corona crisis. *International Journal of Entrepreneurial Behavior & Research, 26*, 1067–1092.

Kraus, S., Kailer, N., Dorfer, J., & Jones, P. (2020b). Open innovation in (young) SMEs. *The International Journal of Entrepreneurship and Innovation, 21*(1), 47–59. https://doi.org/10.1177/1465750319840778

Latifi, M. A., Nikou, S., & Bouwman, H. (2021). Business model innovation and firm performance: Exploring causal mechanisms in SMEs. *Technovation, 107*. https://doi.org/10.1016/j.technovation.2021.102274

Lim, S. L., & Finkelstein, A. (2012). StakeRare: Using social networks and collaborative filtering for large-scale requirements elicitation. *IEEE Transactions on Software Engineering, 38*(3), 707–735.

Lindič, J., & Da Silva, C. M. (2011). Value proposition as a catalyst for a customer focused innovation. *Management Decision, 49*(10), 1694–1708. https://doi.org/10.1108/00251741111183834

Markides, C. (1997). Strategic innovation. *Sloan Management Review, 38*(3), 9–23.

Melander, L. (2019). Customer involvement in product development: Using voice of the customer for innovation and marketing. *Benchmarking: An International Journal, 27*(1), 215–231. https://doi.org/10.1108/BIJ-04-2018-0112

Newbert, S. L., Tornikoski, E. T., & Augugliaro, J. (2020). To get out of the building or not? That is the question: The benefits (and costs) of customer involvement during the startup process. *Journal of Business Venturing Insights, 14.* https://doi.org/10.1016/j.jbvi.2020.e00209

Nooteboom, B. (1994). Innovation and diffusion in small firms: Theory and evidence. *Small Business Economics, 6*(5), 327–347 (1994).

Paternoster, N., Giardino, C., Unterkalmsteiner, M., Gorschek, T., & Abrahamsson, P. (2014). Software development in startup companies: A systematic mapping study. *Information and Software Technology, 56*(10), 1200–1218.

Praag, V. C. M., & Versloot, P. H. (2007). What is the value of entrepreneurship? A review of recent research. *Small Business Economics, 29*(4), 351–382.

Pucihar, A., Lenart, G., Borštnar, M. K., Vidmar, D., & Marolt, M. (2019). Drivers and outcomes of business model innovation-micro, small and medium-sized enterprises perspective. *Sustainability, 2019*(11), 344. https://doi.org/10.3390/su11020344

Rafiq, U., Melegati, J., Khanna, D., Guerra, E., & Wang, X. (2021). Analytics mistakes that derail software startups. In *Evaluation and assessment in software engineering* (pp. 60–69). https://doi.org/10.1145/3463274.3463305

Renzel, D., Behrendt, M., Klamma, R., & Jarke, M. (2013). Requirements bazaar: social requirements engineering for community-driven innovation. In *Proceedings of the RE*, pp. 326–327.

Rintamäki, T., & Saarijärvi, H. (2021). An integrative framework for managing customer value propositions. *Journal of Business Research, 134*, 754–764.

Rosenbusch, N., Brinckmann, J., & Bausch, A. (2011). Is innovation always beneficial? A meta-analysis of the relationship between innovation and performance in SMEs. *Journal of Business Venturing, 26*(4), 441–457. https://doi.org/10.1016/j.jbusvent.2009.12.002

Saad, J., Martinelli, S., Machado, L. S., de Souza, C. R., Alvaro, A., & Zaina, L. (2021). UX work in software startups: A thematic analysis of the literature. *Information and Software Technology, 140*, 106688.

Schmidt, A. L., & Scaringella, L. (2020). Uncovering disruptors' business model innovation activities: Evidencing the relationships between dynamic capabilities and value proposition innovation. *Journal of Engineering and Technology Management, 57*, 101589. https://doi.org/10.1016/j.jengtecman.2020.101589

Snijders, R., Dalpiaz, F., Brinkkemper, S., Hosseini, M., Ali, R., & Ozum, A. (2015). Refine: A gamified platform for participatory requirements engineering. In *Proceedings of the CrowdRE*, pp. 1–6.

Taghizadeh, S. K., Rahman, S. A., & Hossain, M. M. (2018). Knowledge from customer, for customer or about customer: Which triggers innovation capability the most? *Journal of Knowledge Management, 22*(1), 162–182. https://doi.org/10.1108/JKM-12-2016-0548

Teece, D. J. (2018). Business models and dynamic capabilities. *Long Range Planning, 51*(1), 40–49. https://doi.org/10.1016/j.lrp.2017.06.007

Teece, D. J., Pisano, G., & Shuen, A. (1997). Dynamic capabilities and strategic management. *Strategic Management Journal, 18*(7), 509–533.

Unterkalmsteiner, M., Abrahamsson, P., Wang, X., Nguyen-Duc, A., Shah, S., Bajwa, S. S., Edison, H. (2016). Software startups—A research agenda. *E-Informatica Software Engineering Journal, 10*, 89–123.

Vargo, J., & Seville, E. (2011). Crisis strategic planning for SMEs: Finding the silver lining. *International Journal of Production Research, 49*, 5619–5635.

Vossen, R. W. (1998). Relative strengths and weaknesses of small firms in innovation. *International Small Business Journal, 16*(3), 88–94 (1998).

Wijekoon, A., Salunke, S., & Athaide, G. A. (2021). Customer heterogeneity and innovation-based competitive strategy: A review, synthesis, and research agenda. *Journal of Product Innovation Management, 38*(3), 315–333. https://doi.org/10.1111/jpim.12576

Theoretical Background

2

This chapter presents the review of literature based on which research gaps were identified and further research undertaken to fill those gaps. The literature review helped to conceptualise the previous research studies into a framework that provides overall visualisation of the research as disseminated in the book. Based on literature review, the research objectives and research questions are formulated, as highlighted in Chap. 1. The overall research is finally divided into six research studies and research methodology for each research study is decided accordingly, as highlighted in Chap. 3. The literature review is conducted in domain of value proposition innovation in startup context.

2.1 Startup Challenges and Failure Rates

Startups are temporary organisations that are continuously experimenting to identify a business model, which is scalable and repeatable, thereafter they attain higher growth levels and returns. The startups conduct experimentations with markets to test their assumptions about business model elements, for instance, value proposition, customer segments, customer relationships, partnerships etc. The objectives of such experimentations are to explore market to design business model that best aligns with market realities rather merely based on hypobook. For instance, continuous interactions with customers helps startup to test their assumptions, also called hypobook about value proposition, refine existing hypobook, and finally identify the one that matches with customer real needs. A great business idea will alone not be sufficient to result in market success. The innovative value proposition supported by an accurate business model will be the key to success in markets.

© The Author(s), under exclusive license to Springer Nature Switzerland AG 2022
V. Gupta, *Strategic Value Proposition Innovation Management in Software Startups for Sustained Competitive Advantage*, Synthesis Lectures on Technology, Management, & Entrepreneurship, https://doi.org/10.1007/978-3-031-18322-5_2

Designing the business model involves series of experimentations and plenty of changes and even pivots. Once a scalable and repeatable model is designed, the market sales grow up, the model still needs continuous changes, but they are then more incremental. Continuous business model innovation i.e., changes in model that enhances customer value helps startup to sustain competitive advantage. Value proposition however remain the main elements that is changed whenever a business model is innovated.

Continuous value proposition innovation is the key for competitive advantage in continuously fluctuating business environment. Conducting such innovation require startups to invest efforts in exploring markets to acquire market trends. The well-known theories like customer development (Blank, 2013, 2020) and lean startups (Ries, 2011) focuses on customer centric innovations, with focus on interacting with customers to better explore markets to design accurate business models. Lean methodology suggest that startups should build prototypes to acquire customer feedback to improve their business model assumptions based on validated learning using so called build, measure and learn (B-M-l) loop (Ries, 2011). Business model innovations or value proposition innovations are driven by the learning acquired from interacting with market knowledge sources, for instance customers, to validate the assumptions pertaining to business model elements, for instance value proposition. The prototyping has been reported to be a successful learning technique in global markets for business model innovations (Gupta et al., 2021).

Liability of newness and liability of smallness makes it harder for them to invest innovation efforts (Gama et al., 2019; Gimenez-Fernandez et al., 2020). Liability of smallness signifies that the startups has limited resources that hinder their abilities to effectively implement their strategies, for instance innovation (Gimenez-Fernandez et al., 2020). Liability of newness signifies that startup are new firms operating in market and hence lacks the access to critical resources necessary for their survival (Dibrell et al., 2009; Stinchcombe, 1965). In other words, the startups face problems in competing against the well-established firms in the market because of their newly established status (Gimenez-Fernandez et al., 2020). For instance, startups being new in market may have less contacts with customers to involve them in value proposition innovation process which otherwise is not the case with established firms with plenty of royal customers.

Startups are characterized by being innovative, having lack of resources, having more uncertainties, working under time-pressure, having small teams, being highly reactive, and rapidly evolving (Berg et al., 2018). Software startups have challenges which are unique to them. This includes little or no operating history, limited resources, multiple influences from stakeholders, dynamic technologies, and markets (Carmel, 1994). Moreover, they have a less experienced team, high degrees of uncertainties, and tight market release deadlines. They are not self-sustained, highly innovative, highly reactive to changes, and face rapidly scaling requirements (Giardino et al., 2014). These factors hinder the ability of startups to foster innovations.

Despite of these challenges, startups are considered successful innovators (Rosenbusch et al., 2011) although they have high failure rates. One reason could be that market has witnessed many startups with disruptive innovations like Netflix, Uber, Amazon etc. Another reason could be that innovations require firms to respond quickly to changing business environment by seizing opportunities, taking risks, and responding quickly to market feedbacks, which is harder in big companies due to structured processes for everything and hierarchical organizational setups. Startups are agile, have flat organizational setups and quick decision, which makes it possible for them to be highly flexible to foster innovations (Nooteboom, 1994; Rosenbusch et al., 2011; Vossen, 1998). Ability to explore the business environment to design effective business models by overcoming liabilities of smallness and liability of newness will be the key to innovate as well as succeed in the market.

The failures rates of startups are well reported in literature (Mullins et al., 2009; Nobel, 2011; Cantamessa et al., 2018; Danarahmanto et al., 2020; Dvalidze & Markopoulos, 2020; Eesley & Lee, 2020; Haddad et al., 2020; Rafiq et al., 2021; Santisteban et al., 2021). Wrong business model and faulty value proposition, for instance mismatch between product/market fit had been reported as one of the reasons for startup failures (Cantamessa et al., 2018; Danarahmanto et al., 2020; Rafiq et al., 2021). In the coming section a draw is established between value proposition innovation and Requirement engineering as well as software development activities. In literature it had been reported that startups fail because of poor requirement engineering activity, which in management terms signify poor value proposition of product to satisfy customer needs (Alves et al., 2006; Klotins et al., 2015; Giardino et al., 2016; Unterkalmsteiner et al., 2016; Chanin et al., 2017).

Rafiq et al. (2021) reported that the primary reasons for startup failures is related to information collection i.e., either collecting wrong information that leads to nothing, called unproductive information or knowing nothing about what information is to be collected. This pins to activity to elicit market information to formulate winning value propositions.

The startup failure rates are also well reported by the public agencies as well as leading private firms as well. For instance, studies conducted by CB Insights (CB Insights, 2018) U.S. Small Business Administration (SBA) Office of Advocacy (U.S. Small Business Administration, 2018), Statista (Statista, 2020) and Eurostat (Directorate-General of the European Commission) (Eurostat, 2020), also reports that startups have high failure rates, and they cannot survive for long time frames.

2.2 Competitive Advantage

Competitive advantage is the ability of the firms to outperform their competitors in the same industry or markets (Porter, 1985, 1998). In other words, competitive advantage is the edge the firm has over its competitors leading to reduced production costs or better products or services relative to its competitors' offerings. The competitive advantage of a firm is its superior performance relative to its competitors which results in better market share, market share stability, improved business financial position in the market and much more. The firms require sustained competitive advantage. Sustained competitive advantage is its superior performance relative to its competitors over a longer time.

There are two ways of attaining competitive advantage (Porter, 1985). The first one is through cost leadership strategy, which signify that firms reduce its cost of production by optimising its business operations or through economics of scale relative to its competitor offerings. The reduction in cost will reduce the price of the product which enhances the perceived customer value, i.e., the difference between perceived benefits and perceived price. The second way is through differentiation strategy which signify that firms create better products or services relative to competitors. This could include improved performance, better functionality etc.

There could be multiple sources of this competitive advantage, for instance, position of firm in industry (Porter, 1985), resources processed by the firm which are Valuable, Rare, Costly to imitate (inimitable), and Organization (VRIO) (Barney, 1991) and dynamic capabilities processed by the firm (Teece et al., 1997).

The VRIO framework is based on resource-based view of competitive advantage (Barney, 1991), which signifies that the firms attain competitive advantage if they own strategic resources i.e., resources which are Valuable, Rare, Costly to imitate (inimitable) and Organization. These internal competencies of the firm are valuable for them and are harder for their competitors to imitate, which means that these firms have sustained competitive advantage. The competitive advantage is achieved by using these VRIO resources to result in cost reduction or improved differentiation of product or services. However, just procession of the VRIO resources does not guarantee superior performance if these resources are combined in an inefficient manner (Katkalo et al., 2010). Firms should own capability to use these VRIO resources in a way that makes the firm competitively better.

The business environment is changing too fast. These changes require firms to upgrade their competencies and resources continuously to adapt to the environmental changes. This is because environmental changes could require firms to upgrade its resource base resources and the operational competencies to use these resources in a way to adjust to environmental changes, for instance, adopting digital innovation to transform existing operational competencies.

Dynamic capability view is another perspective within the resource-based view (RBV) to explain how firms could have competitive advantage based on their dynamic capabilities. This signify that the firms should continuously refine their existing core competencies; competencies that arise from interplay between resources and competencies, as a response to the environment fluctuations.

Dynamic capabilities theory suggests that the firms are able to make a response to dynamic business environment by renewing their existing competencies through dynamic capabilities (Schriber & Löwstedt, 2020). Firm dynamic capabilities are responsible for business model innovations i.e., their design, implementation, and evolution (Teece, 2018).

Dynamic capabilities of the firm signify their skills or capabilities in responding to fluctuations in business environment by executing three activities-sensing (identifying the market changes in terms of opportunities and threats), seizing (mobilising resources to innovate successfully to turn external turbulence into business favour), and transforming (continuous renewal) (Teece, 2018; Schoemaker et al., 2018).

Thus, the firms should always evolve their operational capabilities by employing their dynamic capabilities. For instance, the software startups could identify arising opportunity in the telemedicine area. To seize this opportunity, they could deploy their existing resources and could collaborate with big software companies to extend their resource base. The use of dynamic capabilities will help the startup to launch innovative product in the telemedicine industry by evolving their existing competencies and resources.

Teece (2018) suggested that business model innovation is triggered by firm's dynamic capabilities. Business model innovation is a way to respond to environmental fluctuations, but this requires strong dynamic capabilities. Dynamic capabilities help firms to implement the business model which is validated for its ability to result in market success. Teece (2018) also suggested that dynamic capabilities help these firms to continuously innovate their business models through three types of dynamic capabilities routines which includesability to identify opportunities (sense), design and refine business model by mobilising resources (seize) and realign structure and culture (transform). Also, organizational flexibility strongly influences its dynamic capabilities to sense, seize and transform market opportunities.

In the software industry, technologies are changing too fast which makes the business environment too dynamic. The business model innovation in general and value proposition in particular in software startup context strongly depend on their dynamic capabilities to adapt to the changing environment and have sustainable competitive advantage.

In highly competitive markets with continuous improvements in innovative product offerings by different firms, sustaining competitive advantage is getting harder for the firms. The small and medium-sized enterprises (SMEs) have resource limitations, especially strategic resources, which means that they will find it harder to achieve and sustain competitive advantage (Fabrizio et al., 2021).

However, organizational flexibility of startups is the main source of their competitive advantage (Fabrizio et al., 2021; Fujita, 2012; Levy & Powell, 1998). Organizational flexibility had been recognised as startups dynamic capabilities (Fabrizio et al., 2021). Flexibility signify that the startups have the ability to make quick decisions to take new courses of actions or change directions as per changing business environment.

As the business environment is changing too fast, the startups could foster innovation to gain competitive advantage if they quickly sense the fluctuations and make corresponding changes in their planned actions. Ability to respond quickly to changes in the business environment is the biggest factor that hinders innovations in big firms.

Innovation had been reported as the factor responsible for competitive advantage (Bashir & Farooq, 2019; Salunke et al., 2019; Distanont & Khongmalai, 2020; Falahat et al., 2020; Barforoush et al., 2021). Business Model Innovation had also been reported positively with sustained competitive advantage (Clauss et al., 2019; Hock-Doepgen et al., 2021; Hossain, 2017; Keiningham et al., 2020).

The dynamic capabilities of startups seem to arise from their flexibilities which in turn fosters business model innovation. The dynamic capabilities theory holds relevance to explore competitive advantage arising from value proposition innovation in software startups as the software industry is changing too fast.

2.3   Value Proposition Innovation as Startup Success Factor

Osterwalder and Pigneur (2010) proposed the business model canvas, a nine element model that capture the value creation, value capture and value delivery aspects of the business. Business model innovation involves creating a new business model from scratch or improving the existing one that enhances customer value. The firms can be more competitive in markets and have higher level of performance if they focus on Business Model Innovation (Vargo & Seville, 2011; Eggers & Kraus, 2011; Kraus et al., 2020a, 2020b; Albats et al., 2021; Clauss et al., 2021; Latifi et al., 2021). The Business Model Innovation is also one of the tools with such firms to make a response to the pandemic (Clauss et al., 2021).

Business model innovation is also the key to creating and capturing the greatest value to be delivered to the customers. The business model innovation requires entrepreneurs to conduct a series of experimentations in the market to test their assumptions about the business model elements. As the experimentations proceed, the business model become more and more accurate and driven by actual market needs rather than based on merely hypobook. One important element of business model is value proposition canvas.

Continuous Value proposition innovation the key element of business model innovation (Morris et al., 2005; Govindarajan & Kopalle, 2006; Chesbrough, 2007; Johnson et al., 2008; Lindič & Da Silva, 2011; Zott et al., 2011; Baden-Fuller & Haefliger, 2013; Bohnsack & Pinkse, 2017). The benefits that the product or service provides to the consumer group are defined by value propositions.

Value proposition is defined "comprehensive description of the value an innovation might offer to an audience, for instance, customer" (Antonopoulou & Begkos, 2020). The value proposition is the bundle of benefits that are offered to the customers. The benefits are offered by launching products or services that makes it possible for customers to use them to do their jobs more efficiently by addressing the problems they were facing in the absence of these products (addressing pains) and finally enhancing the benefits that customers expect from the products (creating gains). In other words, value proposition provides customer value.

The value is considered as the combination of elements that together comprises the customer value (Anderson & Narus, 2006). The customer weights the different elements differently and selects the products that rank higher on the value elements. For instance, value elements could include features, privacy, safety, experience, price and much more. These value elements are derived from the needs of the customers and expectations of benefits. Three types of benefits can be provided to customers namely functional benefits, emotional benefits, and social benefits (Candi & Kahn, 2016). However, the customer value elements are categorised into four categories namely functional, emotional, life changing, and social impact (Almquist et al., 2016). Irrespective of the categories of benefits or customer value elements, they should reduce customer pains and increase expected gains. Startups must identify the elements of value proposition to design solutions that provide innovative value to the customers.

Inventing new value proposition or improving existing one, in other words value proposition innovation, will be the main reason that why customers will prefer to buy the firm product over their competitor's offerings (Johnson et al., 2008; Morris et al., 2005). Value proposition innovation helps startups to identify market opportunities (Guo et al., 2021), create new markets or expand existing ones (Guo et al., 2021), increase customer retention rates (Chesbrough, 2007), improve relationships with customers (Guo et al., 2021) and exploring entrepreneurial opportunities in highly uncertain environments (Chesbrough, 2010; Clauss, 2017; Guo et al., 2021). The business impacts of innovating value proposition includes-improved firm performance (Chandler et al., 2014; Covin et al., 2015)), competitive advantage (Antonopoulou & Begkos, 2020; Rintamäki & Saarijärvi, 2021) and increased revenues (Chesbrough, 2007).

The value proposition innovation is an external dynamic capability of the startup which aim to identify external opportunities and implement them as a response to the continuously changing business environment (Guo et al., 2021; Schmidt & Scaringella, 2020; Teece et al., 1997). Further, value proposition innovation could result in disruptive innovations (Schmidt & Scaringella, 2020). However, value proposition is still an under-developed concept in literature (Antonopoulou & Begkos, 2020) and the research in value proposition innovation domain is limited (Schmidt & Scaringella, 2020).

The product or service value proposition must be able to have the fit with customer needs i.e., should be able to address their problems and provide them with the expected benefits. The previous research indicates that the high failure rates among startups is due to their inability to have a fit between their products and the customer needs. In other words, the value proposition provided by the products does not meet the customer needs i.e., they fail to address their problems and fails to provide them with the expected benefits. Continuous value proposition innovation will be the key for having sustainable competitive advantage leading to improved market position of the startups. Focusing on value proposition innovation will help startups to improve their success rates and grow quickly in the markets.

Considering the importance of this research domain for startups, it is important to empirically analyse the state of art of the value proposition activity to identify future research directions and bridge gaps through more research. The further research will add to the body of knowledge, the knowledge pertaining to value proposition innovation in startups, to help them grow in market and improve the industrial sectors through their innovative product offerings. To achieve this objective, the research question to be explored is formulated as follows:

RQ1. What is the current state of value proposition innovation in startups?

2.4   Value Proposition Innovation and Requirement Engineering: Interdisciplinary Knowledge Transfer

The area of business model innovation is growing quickly (Foss & Seabi, 2016) but the concept of value proposition is still underdeveloped (Antonopoulou & Begkos, 2020) and limited research is conducted in domain of value proposition innovation (Schmidt & Scaringella, 2020). Figure 2.1 shows the four main activities of value proposition innovation.

To the best of our knowledge, the literature lacks the studies that help startups to foster value proposition innovation by explicitly focusing on inventing value proposition ideas and their implementation.

Focusing on inventing value proposition should consider the co-creation with customers and open innovation with other elements of the innovation ecosystem. This help startups to identify innovative value proposition that when implemented and commercialized will help it to grow quickly in the markets.

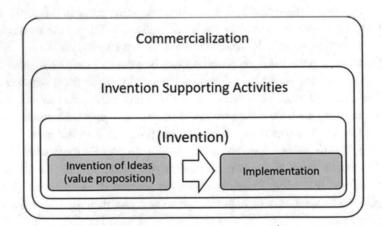

Fig. 2.1   Value Proposition Innovation activities

Focusing on implementation of the innovative value proposition ideas will help startup to undertake the product development process in an innovative way. The process innovations associated with ideas implementation could help to reduce defects, increase quality, reduce time to markets and decrease development costs which could impact the customer perceived value, difference between value and price.

A holistic view of value proposition innovation (Fig. 2.1) is required to foster value proposition innovation, not only focusing on innovative ideas but also their implementation, satisfying the unique constrains in startup context.

As the literature in management sciences pertaining to value proposition innovation especially in startup context is limited, there could be useful lessons to draw from the similar areas in technical sciences like software engineering. For example, the value proposition innovation in the management sciences has a close but unexplored relation with the evolutionary software development in general and requirement engineering activity in particular.

In engineering terms, Requirement engineering deals with identifying innovative value proposition and software development engages in implementing them as working software. The objective is not to provide a technical solution but to draw close analogies between these two similar areas, empirically analyse the technical literature (including the one with interdisciplinary nature, for instance management sciences or social sciences) and draw lessons for improving value proposition innovation from a management perspective.

The value proposition is the combination of benefits or elements that provides value to customers. In software terminologies, the customer needs correspond to user requirements, which may be functional or non-functional. However, non-technical users always specify their pain points and their expectations (in some cases) rather than expressing their software requirements (for software products of software startups). In particular, customer needs are more non-technical and problem domain oriented and software requirements are more technical and solution domain oriented.

The terminological differences between customer needs and product requirements can be ignored as the value proposition and software requirements has same objectives-addressing customer pains (meeting their needs) and providing them the benefits that they expect from using the solution. So, whether the customer specifies their needs in a technical or non-technical way, is immaterial for the research.

In particular, evolutionary software development signifies the innovations in products. The idea generation relates to value proposition identification and software development, for instance, designing, testing, coding and release relates to value proposition innovation. Together, they comprise value proposition innovation, i.e., identification of value proposition, its implementation and commercialisation. The more marketing-oriented commercialisation is outside the scope of this argument. The value proposition could be incremental, such as tiny changes to an existing product, or radical, such as generating wholly new products for the market. Whether incremental or radical, value proposition

Table 2.1 Correspondence between value proposition innovation and evolutionary software engineering

S. No.	Value proposition innovation activity	Evolutionary software development activity	Remarks
1	*Inventing value proposition ideas*	Requirement elicitation (including prototyping for validated learning)	Both aim to identify the needs of customers. The solution will solve customer problems and deliver them benefits
2	*Implementation*	Software development (design, testing, release)	Both deal with implementing ideas into working solutions, for instance, working software
3	*Invention supporting activities* (documentation and prioritization)	*Invention supporting activities* (documentation and prioritization)	Activities deal with supporting invention and implementation activities. For example, documentation and prioritization does not result in new ideas nor implementation

innovation is the key to gaining a competitive edge. The correspondence between value proposition innovation and evolutionary software development as observed through the lens of software development methodology is given in Table 2.1.

By analysing the literature of how software is innovated by continuously gathering new customer requirements (or value propositions) and their implementation (using development models like Agile models), the knowledge extracted from empirical analysing including process models, challenges and observed benefits could be transferred to management literature. This will have two outcomes:

(a) Adding to management literature the value proposition innovation related knowledge. This will help researchers to use the interdisciplinary knowledge to provide good management solutions to entrepreneurs to foster innovations. The knowledge acquired from technical literature will be mapped to management literature by normalising it to non-technical form that aids to direct adoption in management sciences. This help to use the knowledge available in other domains that otherwise remain unutilised due to terminological issues. The empirical analysis of technical literature will only be required for exploration of the value proposition innovation (RQ1 and RQ2). Once the knowledge is acquired and mapped to management domain, the added knowledge will foster further research in management domains about value proposition innovation with support from knowledge acquired from technical literature (RQ3, RQ4, RQ5 and RQ6).

(b) The further research driven by outcomes of RQ1 and RQ2 from management per-
    spective provides good research directions to technical researchers to adopt the
    management learning from technical perspective. This two way knowledge exchange
    will create synergies between management and software engineering. However, the
    scope of this book is limited to management and knowledge transfer to technical
    researchers is merely directions for research community to follow in future.

2.5 Fostering Value Proposition Innovation in Startups Driven by Open Innovation

Business model innovation literature in startup context is still an underdeveloped issue
(Ibarra et al., 2020; Pucihar et al., 2019). More research is required to explore this area to
help startups overcome the barriers to business model innovation. Value proposition innovation being the central element of business model innovation also is an underdeveloped
area especially in startup context.

The barriers in conducting business model innovation in general and value proposition
innovation in particular includes limited resources like financial, R&D facilities, technical
competencies, human resources, less structured processes for innovation and short time
to market (Arbussa et al., 2017; Berends et al., 2014).

As discussed in the preceding section, the business model innovation results in competitive advantage and increased firm performance levels. Business model innovation
involves changes in many canvas elements, especially value proposition canvas. Continuous value proposition innovation is an external dynamic capability of startup to
have sustainable competitive advantage in a fluctuating business environment. For this
to happen, the value proposition helps startup to sense entrepreneurial opportunities in
environment and turn them into valuable products or services.

However, the limited resources owned by startups could inhibit their abilities in searching for opportunities in the environment (Andersen et al., 2022) and hence could limit their
competencies in innovation processes in general (Spender et al., 2017) and value proposition innovation in particular. Further, limited technical capabilities, liability of newness
and limited resources will hinder their abilities to implement the ideas. In particular, liability of smallness and newness hinder their abilities to foster innovations (Gama et al.,
2019; Gimenez-Fernandez et al., 2020).

Adopting open innovation will help startups to overcome their liabilities of newness
and smallness to foster innovations (Hilmersson & Hilmersson, 2021; Kraus et al., 2020a,
2020b; Rosenbusch et al., 2011). The commercialised ideas need to be quickly evolved
as per market learnings i.e., customer feedback.

To overcome these limitations and innovate to maintain competitive advantage, the startups need to implement open innovation by collaborating with open innovation ecosystem elements (Albats et al., 2021; Andersen et al., 2022; Bogers, 2011; Ibarra et al., 2020; Spender et al., 2017). Open innovation is defined as "the use of purposive inflows and outflows of knowledge to accelerate internal innovation, and expand the markets for external use of innovation, respectively" (Chesbrough & Schwartz, 2007). Open innovation involves the two-way exchange of knowledge between the companies and their innovation ecosystem to promote internal innovation to achieve their business objectives. Startups could collaborate with innovation ecosystem elements like academia, customers, freelancers to include them in innovation processes thereby overcoming their liabilities of newness and smallness.

Freelancers could be a good way of getting access to the competencies that startups lack in-house, especially formulating the value proposition. Freelancers could help businesses to find new capabilities, increasing labour force flexibility, increasing speed to market, and innovating new business models (Adamson, 2021). This is in line with the opinion expressed in (Bernabé et al., 2015) about freelancer roles for midsize business. Fuller et al. (2020) conducted a survey with 700 U.S. business leaders associated with firms with revenues greater than 100 million dollars. The survey reported that number of digital platforms for freelancing are increasing, the expected use of such platforms by firms is expected to rise in future and such platforms will be a source of competitive advantage. Further, outsourcing to freelancers has resulted in higher speed to market, increased productivity, and increased innovation in firms. The freelancing could help startups to overcome the barriers to innovation. The growing number of freelancing platforms makes freelancing highly competitive, which makes their prices competitive too. This could be a cost effective option for startups to foster innovation. As value proposition innovation domain is still a developing area and freelancing seems to be a good research direction to explore in startup context as evident from the literature as mentioned before. Motivated by this, this research aims to investigate the answers to the following research questions:

RQ2. What is the current state of freelancer participation in startup value proposition innovation management?

RQ3. What is the current state of practice in startups for freelancer participation in innovation management activities in terms of methodologies, practices, problems, and real-world outcomes?

The objective to investigate empirically freelancer support in value proposition innovation in startup context is to analyse existing research efforts and identification of directions for future research (RQ2). Based on identified future directions, further research efforts will be investigated to explore this domain in real startup settings further (RQ3). In particular, the objective of RQ3 is to explore the freelancer involvement in value proposition innovation in startup context by focusing on strategies for association, challenges incurred and reported business impacts. The outcome will help startup community to adopt the reported outcomes in their business settings to better innovate by overcoming their limitations to innovate. The software development model will be used as a framework to investigate the value proposition innovation activities i.e., invention of ideas and their implementation.

Customers are the main actors of innovation ecosystem. This is because, the startups success in market is dependent if their innovation is going to be adopted by the customers. Value proposition must be aligned with the customer's needs. The customer involvement in value proposition innovation activity will be challenging due to restrictions imposed by the pandemic. This motivated researcher to investigate that how startup could gather ideas from customers to continuously foster value proposition innovation using a variety of technologies available in the marketplace. The role of customers in fostering innovation had already been reported in (Desouza et al., 2008; Lindič & Silva, 2011; Melander, 2019; Taghizadeh et al., 2018; Wijekoon et al., 2021). The research aims to find answer to the following research question.

RQ4. How can value proposition innovation be fostered through customer participation?

The freelancing platforms or technologies are growing and demand across them are expected to grow further (Fuller et al., 2020). Many freelancing platforms are now listed on stock exchanges through initial public offering (Stephany et al., 2021) which signifies that company meet requirements to be listed on stock exchange, has great financial strength, and will further strengthen its share liquidity. Freelancing platforms offer startups the opportunity to access global workforce (de Peuter, 2011), lower their cost of labour (de Peuter, 2011; Popiel, 2017) and increase company financial valuations (Irani, 2015). This led to motivation to investigate the better ways of strategically outsourcing to freelancers, by making the best selection cross the available freelancing platforms. The research aims to find answer to the following research question.

RQ5. How can entrepreneurs assess freelancing platforms to create and sustain strategic relationships with freelancers?

The importance of academia as one of the innovation ecosystem elements for the business had been reported in (Mehta, 2004; Münch et al., 2013; Saguy, 2016). The academia can help startups in value proposition innovation by providing access to resources for generating new ideas, implementing them as well as commercialising them. The research aims to find answer to the following research question.

RQ6. How might academic strategic partnerships be investigated as a means of promoting innovation management?

The literature provided directions for the research that forms part of this book. It helped to provide a basis for the research questions that this book tries to answer. Analysis of previous research studies was necessary to identify research gaps that justifies the investing of further research gaps. Figure 2.2 represent the research gaps and areas focused by the book.

Overall, this helped to conceptualise the research into the conceptual framework as given in Fig. 2.3.

The conceptual framework illustrates that the research study will involve investigating state of art and state of practice. State of art involves exploring research trends in value proposition innovation and freelancer supported value proposition innovation. State of practice aims to explore the freelancer, customer and academia supported value proposition innovation. This involves exploring the real practices of startups by investigating their methods of associating with freelancers, involved challenges and reported business impacts. Further, the innovations in technology like freelancer platforms and feedback

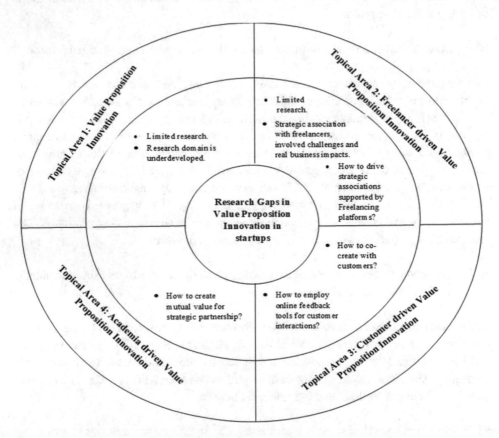

Fig. 2.2 Research gaps and research areas

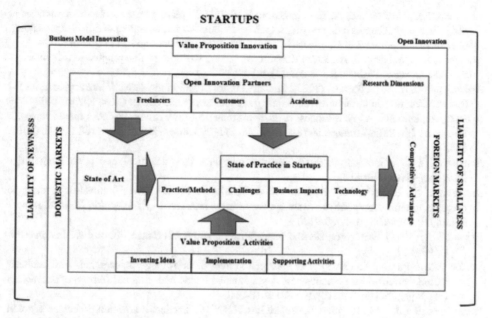

Fig. 2.3 Conceptual framework

management systems could be further explored to evaluate the promising technologies and integrate them to seamlessly integrate freelancers and customers in value proposition innovation activities. The strategic partnerships with academia could also help startups to foster value proposition innovation. The overall impact of research will be a contribution to the startup community in increasing their competitive advantage which finally leads to market success. The dynamic competencies theory and resource based-view are used to explore if freelancers, customers, and academia support could help startups to build their dynamic competencies to undertake value proposition innovation and finally leading to sustainable competitive advantage.

References

Adamson, C. (2021). How freelance developers can help your midsize business. *Harvard Business Review*.

Albats, E., Podmetina, D., & Vanhaverbeke, W. (2021). Open innovation in SMEs: A process view towards business model innovation. *Journal of Small Business Management*, 1–42. https://doi.org/10.1080/00472778.2021.1913595

Almquist, E., Senior, J., & Bloch, N. (2016). The element of value: measuring and delivering what consumers really want. *Harvard Business Review*.

Alves, C., Pereira, S., & Castro, J. (2006). A study in market-driven requirements engineering. In: *Proceedings of the 9th Workshop on Requirements Engineering (WER '06)*, Rio de Janeiro, Brazil.

Andersen, T. C. K., Aagaard, A., & Magnusson, M. (2022). Exploring business model innovation in SMEs in a digital context: Organizing search behaviours, experimentation and decision-making. *Creativity and Innovation Management, 31*(1), 19–34. https://doi.org/10.1111/caim.12474

Anderson, J. C., & Narus, J. A., & van Rossum, W. (2006). Customer value propositions in business markets. *Harvard Business Review, 84*(3), 90–99.

Antonopoulou, K., & Begkos, C. (2020). Strategizing for digital innovations: Value propositions for transcending market boundaries. *Technological Forecasting and Social Change, 156*, 120042.

Arbussa, A., Bikfalvi, A., & Marquès, P. (2017). Strategic agility-driven business model renewal: The case of an SME. *Management Decision, 55*, 271–293. https://doi.org/10.1108/MD-05-2016-0355

Baden-Fuller, C., & Haefliger, S. (2013). Business models and technological innovation. *Long Range Planning, 46*(6), 419–426.

Barforoush, N., Etebarian, A., Naghsh, A., & Shahin, A. (2021). Green innovation a strategic resource to attain competitive advantage. *International Journal of Innovation Science*. https://doi.org/10.1108/IJIS-10-2020-0180

Barney, J. B. (1991). Firm resources and sustained competitive advantage. *Journal of Management, 17*(1), 99–120.

Bashir, M., & Farooq, R. (2019). The synergetic effect of knowledge management and business model innovation on firm competence: A systematic review. *International Journal of Innovation Science*. https://doi.org/10.1108/IJIS-10-2018-0103

Berends, H., Jelinek, M., Reymen, I., & Stultiëns, R. (2014). Product innovation processes in small firms: Combining entrepreneurial effectuation and managerial causation. *Journal of Product Innovation Management, 31*, 616–635. https://doi.org/10.1111/jpim.12117

Berg, V., Birkeland, J., Nguyen-Duc, A., Pappas, I. O., & Jaccheri, L. (2018). Software startup engineering: A systematic mapping study. *Journal of Systems and Software, 144*, 255–274.

Bernabé, R. B., Navia, I.Á., & García-Peñalvo, F. J. (2015). Faat: Freelance as a team. In *Proceedings of the 3rd International Conference on Technological Ecosystems for Enhancing Multiculturality*, Porto, Portugal, pp. 687–694.

Blank, S. (2013). Why the lean start-up changes everything. *Harvard Business Review, 91*(5), 63–72.

Blank, S. (2020). *The four steps to the epiphany: Successful strategies for products that win*. Wiley.

Bogers, M. (2011). The open innovation paradox: Knowledge sharing and protection in R&D collaborations. *European Journal of Innovation Management, 14*(1), 93–117.

Bohnsack, R., & Pinkse, J. (2017). Value propositions for disruptive technologies: Reconfiguration tactics in the case of electric vehicles. *California Management Review, 59*(4), 79–96. https://doi.org/10.1177/0008125617717711

Candi, M., & Kahn, K. B. (2016). Functional, emotional, and social benefits of new B2B services. *Industrial Marketing Management, 57*, 177–184. https://doi.org/10.1016/j.indmarman.2016.02.002

Cantamessa, M., Gatteschi, V., Perboli, G., & Rosano, M. (2018). Startups' roads to failure. *Sustainability, 10*(7), 2346. https://doi.org/10.3390/su10072346

Carmel, E. (1994). Time-to-completion in software package startups. In *Proceedings of the 27th Hawaii International Conference on System Sciences (HICSS)*, Maui, HI, USA, Vol. 4, No. 7, pp. 498–507.

CB Insights. (2018). *CB insights venture capital funnel shows odds of becoming a unicorn are about 1%*. Retrieved 10 March 2021, from https://www.statista.com/statistics/1114070/eu-business-survival-rates-by-country-2017/

Chandler, G. N., Broberg, J. C., & Allison, T. H. (2014). Customer value propositions in declining industries: Differences between industry representative and high-growth firms. *Strategic Entrepreneurship Journal, 8*(3), 234–253.

Chanin, R., Pompermaier, L., Fraga, K., Sales, A., Prikladnicki, R. (2017). Applying customer development for software requirements in a startup development program. In *Proceedings of the 2017 IEEE/ACM 1st International Workshop on Software Engineering for Startups (SoftStart)*, Buenos Aires, Argentina, 21 May 2017, pp. 2–5.

Chesbrough, H. (2007). Business model innovation: It's not just about technology anymore. *Strategy & Leadership, 35*(6), 12–17.

Chesbrough, H. (2010). Business model innovation: Opportunities and barriers. *Long Range Planning, 43*(2), 354–363.

Chesbrough, H., & Schwartz, K. (2007). Innovating business models with co-development partnerships. *Research-Technology Management, 50*(1), 55–59.

Clauss, T. (2017). Measuring business model innovation: Conceptualization, scale development, and proof of performance. *R&d Management, 47*(3), 385–403.

Clauss, T., Abebe, M., Tangpong, C., & Hock, M. (2019). Strategic agility, business model innovation, and firm performance: An empirical investigation. *IEEE Transactions on Engineering Management, 68*(3), 767–784. https://doi.org/10.1109/TEM.2019.2910381

Clauss, T., Breier, M., Kraus, S., Durst, S., & Mahto, R. V. (2021). Temporary business model innovation–SMEs' innovation response to the Covid-19 crisis. *R&D Management*. https://doi.org/10.1111/radm.12498

Covin, J. G., Garrett, R. P., Jr., Kuratko, D. F., & Shepherd, D. A. (2015). Value proposition evolution and the performance of internal corporate ventures. *Journal of Business Venturing, 30*(5), 749–774.

Danarahmanto, P. A., Primiana, I., Azis, Y., & Kaltum, U. (2020). The sustainable performance of the digital start-up company based on customer participation, innovation, and business model. *Business: Theory and Practice, 21*(1), 115–124.

de Peuter, G. (2011). Creative economy and labor precarity: A contested convergence. *Journal of Communication Inquiry, 35*(4), 417–425. https://doi.org/10.1177/0196859911416362

Desouza, K. C., Awazu, Y., Jha, S., Dombrowski, C., Papagari, S., Baloh, P., & Kim, J. Y. (2008). Customer-Driven Innovation. *Research-Technology Management, 51*(3), 35–44. https://doi.org/10.1080/08956308.2008.11657503

Dibrell, C., Craig, J. B., Moores, K., Johnson, A. J., & Davis, P. S. (2009). Factors critical in overcoming the liability of newness: Highlighting the role of family. *The Journal of Private Equity, 12*(2), 38–48. http://www.jstor.org/stable/43503588

Distanont, A., & Khongmalai, O. (2020). The role of innovation in creating a competitive advantage. *Kasetsart Journal of Social Sciences, 41*(1), 15–21.

Dvalidze, N., & Markopoulos, E. (2019). Understanding the nature of entrepreneurial leadership in the startups across the stages of the startup lifecycle. In *International Conference on Applied Human Factors and Ergonomics* (pp. 281–292). Springer, Cham.

Eesley, C. E., & Lee, Y. S. (2020). Do university entrepreneurship programs promote entrepreneurship? *Strategic Management Journal, 42*(4), 833–861. https://doi.org/10.1002/smj.3246

Eggers, F., & Kraus, S. (2011). Growing young SMEs in hard economic times: The impact of entrepreneurial and customer orientations—A qualitative study from Silicon Valley. *Journal of Small Business & Entrepreneurship, 24*, 99–111.

Eurostat. (2020). *Business demography statistics*. Retrieved 26 March 2021, from https://ec.eur opa.eu/eurostat/statistics-explained/index.php/Business_demography_statistics#Enterprise_surv ival_rate

Fabrizio, C. M., Kaczam, F., de Moura, G. L., da Silva, L. S. C. V., da Silva, W. V., & da Veiga, C. P. (2021). Competitive advantage and dynamic capability in small and medium-sized enterprises: A systematic literature review and future research directions. *Review of Managerial Science*, 1–32. https://doi.org/10.1007/s11846-021-00459-8

Falahat, M., Ramayah, T., Soto-Acosta, P., & Lee, Y. Y. (2020). SMEs internationalization: The role of product innovation, market intelligence, pricing and marketing communication capabilities as drivers of SMEs' international performance. *Technological Forecasting and Social Change, 152*. https://doi.org/10.1016/j.techfore.2020.119908

Foss, N. J., & Saebi, T. (2017). Fifteen years of research on business model innovation: How far have we come, and where should we go? *Journal of Management, 43*(1), 200–227. https://doi.org/10.1177/0149206316675927

Fujita, M. (2012). *The transnational activities of small and medium-sized enterprises.* Springer Science & Business Media.

Fuller, J. B., Raman, M., Palano, J., Bailey, A., Vaduganathan, N., Kaufman, E., Laverdière, R., & Lovett, S. (2020). *Building the on-demand workforce.* Harvard Business School and BCG.

Gama, F., Frishammar, J., & Parida, V. (2019). Idea generation and open innovation in SMEs: When does market-based collaboration pay off most? *Creativity and Innovation Management, 28*(1), 113–123. https://doi.org/10.1111/caim.12274

Giardino, C., Unterkalmsteiner, M., Paternoster, N., Gorschek, T., & Abrahamsson, P. (2014). What do we know about software development in startups? *IEEE Software, 31*(5), 28–32.

Giardino, C., Paternoster, N., Unterkalmsteiner, M., Gorschek, T., & Abrahamsson, P. (2016). Software development in startup companies: The greenfield startup model. *IEEE Transactions on Software Engineering, 42*, 585–604.

Gimenez-Fernandez, E. M., Sandulli, F. D., & Bogers, M. (2020). Unpacking liabilities of newness and smallness in innovative start-ups: Investigating the differences in innovation performance between new and older small firms. *Research Policy, 49*(10). https://doi.org/10.1016/j.respol.2020.104049.

Govindarajan, V., & Kopalle, P. K. (2006). The usefulness of measuring disruptiveness of innovations ex post in making ex ante predictions. *Journal of Product Innovation Management, 23*(1), 12–18.

Guo, H., Yang, J., & Han, J. (2021). The fit between value proposition innovation and technological innovation in the digital environment: Implications for the performance of startups. *IEEE Transactions on Engineering Management, 68*(3), 797–809. https://doi.org/10.1109/TEM.2019.2918931

Gupta, V., Rubalcaba, L., & Gupta, C. (2021). Multimedia prototyping for early-stage startups endurance: Stage for new normal? *IEEE Multimedia, 28*(4), 107–116.

Haddad, H., Weking, J., Hermes, S., Böhm, M., & Krcmar, H. (2020). Business model choice matters: How business models impact different performance measures of startups. In *Wirtschaftsinformatik (Zentrale Tracks)* (pp. 828–843).

Hilmersson, F. P., & Hilmersson, M. (2021). Networking to accelerate the pace of SME innovations. *Journal of Innovation & Knowledge, 6*(1), 43–49. https://doi.org/10.1016/j.jik.2020.10.001

Hock-Doepgen, M., Clauss, T., Kraus, S., & Cheng, C. F. (2021). Knowledge management capabilities and organizational risk-taking for business model innovation in SMEs. *Journal of Business Research, 130*, 683–697. https://doi.org/10.1016/j.jbusres.2019.12.001

Hossain, M. (2017). Business model innovation: Past research, current debates, and future directions. *Journal of Strategy and Management, 10*(3), 342–359. https://doi.org/10.1108/JSMA-01-2016-0002

Ibarra, D., Bigdeli, A. Z., Igartua, J. I., & Ganzarain, J. (2020). Business model innovation in established SMEs: A configurational approach. *Journal of Open Innovation: Technology, Market, and Complexity, 6*(3), 76. https://doi.org/10.3390/joitmc6030076

Irani, L. (2015). Difference and dependence among digital workers: The case of amazon mechanical turk. *South Atlantic Quarterly, 114*(1), 225–234. https://doi.org/10.1215/00382876-2831665

Johnson, M. W., Christensen, C. M., & Kagermann, H. (2008). Reinventing your business model. *Harvard Business Review, 86*(12), 57–68.

Katkalo, V., Pitelis, C., & Teece, D. (2010). Introduction: On the nature and scope of dynamic capabilities. *Industrial and Corporate Change, 19*(4), 1175–1186.

Keiningham, T., Aksoy, L., Bruce, H. L., Cadet, F., Clennell, N., Hodgkinson, I. R., & Kearney, T. (2020). Customer experience driven business model innovation. *Journal of Business Research, 116*, 431–440. https://doi.org/10.1108/JSMA-01-2016-0002

Klotins, E., Unterkalmsteiner, M., & Gorschek, T. (2015). Software engineering knowledge areas in startup companies: A mapping study. In *Proceedings of the International Conference of Software Business*, Braga, Portugal, pp. 245–257.

Kraus, S., Clauss, T., Breier, M., Gast, J., Zardini, A., & Tiberius, V. (2020a). The economics of COVID-19: Initial empirical evidence on how family firms in five European countries cope with the corona crisis. *International Journal of Entrepreneurial Behavior & Research, 26*, 1067–1092.

Kraus, S., Kailer, N., Dorfer, J., & Jones, P. (2020b). Open innovation in (young) SMEs. *The International Journal of Entrepreneurship and Innovation, 21*(1), 47–59. https://doi.org/10.1177/146 5750319840778

Latifi, M. A., Nikou, S., & Bouwman, H. (2021). Business model innovation and firm performance: Exploring causal mechanisms in SMEs. *Technovation, 107*. https://doi.org/10.1016/j.technovat ion.2021.102274

Levy, M., & Powell, P. (1998). SME flexibility and the role of information systems. *Small Business Economics, 11*(2), 183–196.

Lindič, J., & Da Silva, C. M. (2011). Value proposition as a catalyst for a customer focused innovation. *Management Decision, 49*(10), 1694–1708. https://doi.org/10.1108/00251741111183834

Mehta, S. (2004). The emerging role of academia in commercializing innovation. *Nature Biotechnology, 22*, 21–24. https://doi.org/10.1038/nbt0104-21

Melander, L. (2019). Customer involvement in product development: Using voice of the customer for innovation and marketing. *Benchmarking: An International Journal, 27*(1), 215–231. https:// doi.org/10.1108/BIJ-04-2018-0112

Morris, M., Schindehutte, M., & Allen, J. (2005). The entrepreneur's business model: Toward a unified perspective. *Journal of Business Research, 58*(6), 726–735.

Mullins, J., Mullins, J. W., Mullins, J. W., & Komisar, R. (2009). *Getting to plan B: Breaking through to a better business model.* Harvard Business Press.

Münch, J., Fagerholm, F., Johnson, P., Pirttilahti, J., Torkkel, J., & Jäarvinen, J. (2013). Creating minimum viable products in industry academia collaborations. In *International Conference on Lean Enterprise Software and Systems* (pp. 137–151). Springer. https://doi.org/10.1007/978-3-642-44930-7_9

Nobel, C. (2011). *Teaching a 'lean startup' strategy.* HBS Working Knowledge, pp. 1–2.

Nooteboom, B. (1994). Innovation and diffusion in small firms: Theory and evidence. *Small Business Economics, 6*(5), 327–347.

Osterwalder, A., & Pigneur, Y. (2010). *Business model generation: a handbook for visionaries, game changers, and challengers* (Vol. 1). Wiley. ISBN: 978-0-470-87641-1.

Popiel, P. (2017). "Boundaryless" in the creative economy: Assessing freelancing on Upwork. *Critical Studies in Media Communication, 34*(3), 220–233. https://doi.org/10.1080/15295036.2017. 1282618

Porter, M. E. (1998). *The competitive advantage of the nation.* Palgrave.

Porter, M. E. (1985). *Competitive advantage: Creating and sustaining superior performance.* Free Press.

Pucihar, A., Lenart, G., Borštnar, M. K., Vidmar, D., & Marolt, M. (2019). Drivers and outcomes of business model innovation-micro, small and medium-sized enterprises perspective. *Sustainability, 2019*(11), 344. https://doi.org/10.3390/su11020344

Rafiq, U., Melegati, J., Khanna, D., Guerra, E., & Wang, X. (2021). *Analytics mistakes that derail software startups. in evaluation and assessment in software engineering* (pp. 60–69). https://doi.org/10.1145/3463274.3463305

Ries, E. (2011). *The lean startup: How today's entrepreneurs use continuous innovation to create radically successful businesses.* Currency.

Rintamäki, T., & Saarijärvi, H. (2021). An integrative framework for managing customer value propositions. *Journal of Business Research, 134*, 754–764.

Rosenbusch, N., Brinckmann, J., & Bausch, A. (2011). Is innovation always beneficial? A meta-analysis of the relationship between innovation and performance in SMEs. *Journal of Business Venturing, 26*(4), 441–457. https://doi.org/10.1016/j.jbusvent.2009.12.002

Saguy, I. S. (2016). Challenges and opportunities in food engineering: Modeling, virtualization, open innovation and social responsibility. *Journal of Food Engineering, 176*, 2–8. https://doi.org/10.1016/j.jfoodeng.2015.07.012

Salunke, S., Weerawardena, J., & McColl-Kennedy, J. R. (2019). The central role of knowledge integration capability in service innovation-based competitive strategy. *Industrial Marketing Management, 76*, 144–156. https://doi.org/10.1016/j.indmarman.2018.07.004

Santisteban, J., Mauricio, D., & Cachay, O. (2021). Critical success factors for technology-based startups. *International Journal of Entrepreneurship and Small Business, 42*(4), 397–421. https://doi.org/10.1504/IJESB.2021.114266

Schmidt, A. L., & Scaringella, L. (2020). Uncovering disruptors' business model innovation activities: Evidencing the relationships between dynamic capabilities and value proposition innovation. *Journal of Engineering and Technology Management, 57*, 101589. https://doi.org/10.1016/j.jengtecman.2020.101589

Schoemaker, P. J., Heaton, S., & Teece, D. (2018). Innovation, dynamic capabilities, and leadership. *California Management Review, 61*(1), 15–42. https://doi.org/10.1177/2F0008125618790246

Schriber, S., & Löwstedt, J. (2020). Reconsidering ordinary and dynamic capabilities in strategic change. *European Management Journal, 38*(3), 377–387. https://doi.org/10.1016/j.emj.2019.12.006

Spender, J.-C., Corvello, V., Grimaldi, M., & Rippa, P. (2017). Startups and open innovation: A review of the literature. *European Journal of Innovation Management, 20*(1), 4–30. https://doi.org/10.1108/EJIM-12-2015-0131

Statista. (2020). *Satista business survival rates in selected European countries in 2017, by length of survival.* Retrieved 10 March 2021, from https://www.statista.com/statistics/1114070/eu-business-survival-rates-by-country-2017/

Stephany, F., Kässi, O., Rani, U., & Lehdonvirta, V. (2021). Online Labour Index 2020: New ways to measure the world's remote freelancing market. *Big Data & Society, 8*(2). https://doi.org/10.1177/20539517211043240

Stinchcombe, A. L. (1965). *Organizations and social structure handbook of organizations* (Vol. 44, pp. 142–193).

Taghizadeh, S. K., Rahman, S. A., & Hossain, M. M. (2018). Knowledge from customer, for customer or about customer: Which triggers innovation capability the most? *Journal of Knowledge Management, 22*(1), 162–182. https://doi.org/10.1108/JKM-12-2016-0548

Teece, D. J. (2018). Business models and dynamic capabilities. *Long Range Planning, 51*(1), 40–49. https://doi.org/10.1016/j.lrp.2017.06.007

Teece, D. J., Pisano, G., & Shuen, A. (1997). Dynamic capabilities and strategic management. *Strategic Management Journal, 18*(7), 509–533.

U.S. Small Business Administration. (2018). *Frequently asked questions about small business*. Retrieved 24 March 2021, from https://www.sba.gov/sites/default/files/advocacy/Frequently-Asked-Questions-Small-Business-2018.pdf

Unterkalmsteiner, M., Abrahamsson, P., Wang, X., Nguyen-Duc, A., Shah, S., Bajwa, S. S., Edison, H. (2016). Software startups—A research agenda. *E-Informatica Software Engineering Journal, 10*, 89–123.

Vargo, J., & Seville, E. (2011). Crisis strategic planning for SMEs: Finding the silver lining. *International Journal of Production Research, 49*, 5619–5635.

Vossen, R. W. (1998). Relative strengths and weaknesses of small firms in innovation. *International Small Business Journal, 16*(3), 88–94.

Wijekoon, A., Salunke, S., & Athaide, G. A. (2021). Customer heterogeneity and innovation-based competitive strategy: A review, synthesis, and research agenda. *Journal of Product Innovation Management, 38*(3), 315–333. https://doi.org/10.1111/jpim.12576

Zott, C., Amit, R., & Massa, L. (2011). The business model: Recent developments and future research. *Journal of Management, 37*(4) (2011), 1019–1042. https://doi.org/10.1177/0149206311406265

Research Design

3

This chapter of the book aims to provide insights into multiple research methods employed to meet formulated research objectives. First, generic overview of research methodology adopted in conducting research is specified. After that, focus is made on research methods employed to conduct the research. This involves systematic literature reviews (mappings), case studies, surveys, experience reports and their combination. The guidelines employed for executing various research methods in the project are also presented. Finally, research methodologies adopted in meeting research objectives are highlighted by focusing on individual research studies conducted and disseminated as nine research papers.

3.1 Research Methodology

The sample for research includes the freelancers and the software startups based on Asia and Europe. Most of data collected is quantitative in nature and hence is analysed using grounded theory. Part of quantitative data is analysed using descriptive statistics. This research includes the different research types—systematic literature reviews (mappings), case studies, surveys, experience reports and their combination. The underlying research protocol which defines sampling, data collection, data analysis and result valida-tion considers various validity and ethical issues. The objective is to achieve the research objectives in the most trustworthy manner, satisfying the ethical issues involved in the research. For instance, in case study, data instruments were made aware of Informed con-sent, Review board approval, Confidentiality, Handling of sensitive results, Inducements and Feedback. Member checking is done to access the result validity.

© The Author(s), under exclusive license to Springer Nature Switzerland AG 2022   39
V. Gupta, *Strategic Value Proposition Innovation Management in Software Startups for Sustained Competitive Advantage*, Synthesis Lectures on Technology, Management, & Entrepreneurship, https://doi.org/10.1007/978-3-031-18322-5_3

3.2   Research Methods Employed

The research study is decomposed into six independent but connected research studies, as mentioned in Chap. 1. Each research study addresses a separate research goal, which when combined, achieves the book's overall research goal. Each research study involves different research methods because they have different objectives which could be best met with a particular research method. For instance, synthesising the body of literature could be best achieved with systematic literature surveys. Investigating freelancers' participation in innovation in the real setting of startups could be best explored with case studies due to exploratory and/or descriptive nature of research. As a result, the research methodology includes systematic literature reviews, case studies, surveys, and experience reports.

Table 3.1 shows the relationship between research studies, research questions addressed by research studies, and research methods used to answer individual research questions to achieve the overall research goal.

Next, the different research methods are described along with the empirical guidelines that are executed to conduct the research.

Table 3.1 Research method employed in research article

S. no.	Research study	Research question	Research method
1.	Study 1	What is the current state of value proposition innovation in startups?	Systematic literature reviews
2.	Study 2	What is the current state of freelancer participation in startup innovation management activities?	
3.	Study 3	What is the current state of practice in startups for freelancer participation in innovation management activities in terms of methodologies, practices, problems, and real-world outcomes?	Case study
4.	Study 4	How can value proposition innovation be fostered through customer participation?	Experience reports, case study and surveys
5.	Study 5	How can entrepreneurs assess freelancing platforms to create and sustain strategic relationships with freelancers?	Experience reports, case study and surveys
6.	Study 6	How might academic strategic partnerships be investigated as a means of promoting innovation management?	Reviews and experience reports

3.2.1   Systematic Mapping Study

One of the types of systematic literature reviews is systematic mapping studies. They are aiming to organize the study area into categories (classification schema) that represent the area's research trends and focus. These investigations are carried out to acquire a wide picture of the research available in a given field by classifying; for instance, identifying categories and relationships and counting, for instance, number of publications within classification scheme. This allows researchers to get a comprehensive view of research efforts made by other researchers in the field to uncover research gaps.

The research topics are broad since this study takes a comprehensive view of the subject area (mappings are broad studies rather deeply conducted). These studies are primarily concerned with highlighting research trends rather than providing empirical evidence through in-depth analysis of research findings. As a result, the number of studies that make up a systematic study is typically quite high, and it is synthesized by analysing their abstracts (and conclusion if abstracts are ambiguous). Because of the bias involved, the reliability of such studies is a major concern (Petersen & Gencel, 2013).

In software engineering, the guidelines for conducting the systematic mapping process are reported (Petersen et al., 2008). The updated guidelines are based on a study of prior mapping guidelines and mapping studies that have already been completed are reported in (Petersen et al., 2015). To achieve a higher level of validity in the mapping study, the mapping involves identification of previous research studies (search, inclusion, and exclusion), categorization and classification schemes and processes, and different ways of visualizing the results. The following processes are included in the overall guidelines: definition of research questions (defining research scope), conduct search, screening of the papers, keywords using abstracts and data extraction and mapping process.

A huge number of papers are found after bibliographic databases are examined (performing a search) to answer the broadly formulated research questions. To filter out the non-relevant publications, these papers are processed according to the conditions provided out in a mapping methodology (inclusion and exclusion criteria). To build the classification schema, the pertinent ones are submitted to further investigation. The categorization schema is populated with the information taken from the abstracts (and conclusion). As a result, the mapping process is divided into three stages: planning the mapping protocol, executing the protocol, and reporting the results.

3.2.2   Case Study

Runeson and Höst (2009) written procedures for conducting and reporting software engineering case studies. Case study research consists of five steps: case study design, data gathering procedures, evidence collection, data analysis, and reporting. The case study protocol is identified as a result of the case study design. The protocol includes case selection techniques, data collection procedures, analysis procedures, validity, ethical issues, and other case study features as planned for the case study.

Data collecting methods are used to gather information from a variety of sources, which is then evaluated using a variety of approaches, resulting in the creation and testing of a hypobook. The data is mainly qualitative, but limited quantitative data can also be collected and analysed using descriptive statistics.

The case study's findings are subsequently presented to the target audience using appropriate reporting formats. The case study is carried out with the validity and ethical considerations in mind. The research is carried out in the most reliable manner possible in order to accomplish the stated research objectives while also addressing the ethical concerns raised by the study. Various checklists are offered to aid the researcher in doing the case study in an effective and efficient manner.

3.2.3   Experience Reports

Experience reports are the means of sharing the practical industrial experience with the research community. These reports about value proposition innovation mention the problems faced in this activity, solutions that are employed by startups and implications for their adoption. The mistakes made in the value proposition innovation task could also be a meaningful way to help startup community to avoid those costly mistakes and bring out innovative ways of undertaking innovation. The success stories of startups in fostering value proposition innovation becomes the basis for developing more trust in the report.

3.3   Research Methods in Individual Articles

This section discusses the research methods used to accomplish the research objectives of various research studies in order to meet the overall book objectives. There are one or more research outcomes for each research study. For instance, research study 3 had three research outcomes. The study outcomes, with each outcome distributed in venues such as journals and conferences/workshops (refer to Table 4.1). The Appendix contains the details of the research studies and their outcomes. Refer to Table 1.2 for information on research publications, Appendix for their specifics and Table 4.1 for a concise overview of their correspondence with research papers, research issues, and research methods.

3.3.1   Systematic Mappings Study

Research study 1 aims to investigate the value proposition innovation in startup context to analyse research trends across the area. The objective was not to synthesise the literature findings empirically but to provide a holistic view of research efforts invested in this research domain. The outcome will provide motivation and directions to undertake future research in this domain. Driven by the outcome, Research study 2 aims to analyse research trends across freelancer driven value proposition innovation. The findings will identify current research efforts as well as research topics being investigated along with associated challenges. To meet the objective of investigating the invested research in the area, the Systematic Literature Mappings as the part of systematic literature reviews were conducted. The research was conducted in a way to ensure that four types of threat to validity are addressed, which includes descriptive validity, theoretical validity, generalizability, interpretive validity.

Research Study 1: Systematic Literature Mapping Protocol
The search process involves the formulation of a reliable mapping protocol. The systematic mapping protocol was drafted through various interactive steps involving discussions and negotiations with other researchers. The process of formulation of mapping protocol involves defining research objectives and questions, choosing search strategy, developing search string, evaluating search, defining inclusion and exclusion criteria, extraction and classification strategy and addressing threats to validity. In this research study, the search strategy selected was database search and snowballing rather manually selecting articles to meet research objectives. Search string was developed using the Population, Intervention, Comparison and Outcomes (PICO) framework. The extraction and classification strategy involved both topic-independent and topic-dependent classifications. Data extraction form was used to capture the required information from the research studies accurately.

The search strategy was validated by comparing the selected research studies to well-known articles published in leading venues about requirement engineering in the context of software startups. The neutrality of the inclusion and exclusion criteria is ensured by running them through the same bibliographic databases with the same search string, but only for the year 2020. The researchers' chosen publications are then compared to see how well the picks coincide. Small conflicts were settled through consensus meetings, and some criteria for inclusion and exclusion were revised to make them more objective.

The mapping technique is carried out by triggering the bibliographic database using the search string entered, resulting in the initial list of studies. The initial list of retrieved research from four bibliographic databases was filtered numerous times against inclusion and exclusion criteria in order to reduce the number of studies to the "most suited." The actions listed below are part of the various stages:

- The studies are analyzed in two stages: first on the basis of their titles, and subsequently on the basis of their abstracts.
- A third round of screening is performed on the studies identified in the previous step. Their Google citations are extracted at this point, and they are screened by reading their title and abstract (the forward snowballing approach provided in Wohlin (2014).
- The studies identified in previous steps are the ones that will be analysed to meet achieve research objectives.

The research studies that passed the inclusion criteria were subjected to analysis as per data aggregation and synthesis approach planned in the protocol planning phase. The findings are then reported in the form of research article (Article 1, refer to Appendix I).

Article 2: Systematic Literature Mapping Protocol
The research study involves the same systematic mapping protocol that was employed in study 1. The mapping protocol was executed, and findings reported in Article 2. Employing of multiple researchers in the process of planning and executing mapping protocol ensures the minimisation of bias and more objectivity in the research process. Random checks were also performed to cross-check the information gathered from the mapping protocol execution. The mapping protocol's conception and execution were broken down into a series of milestones. Before going on to greater goals, each milestone is checked and agreed upon as it is reached. For instance, sub schemas were constructed and considered as small milestones before constructing a big structure schema for each research questions. The milestones were validated by all of the authors. Last but not least, random efforts were made to review all milestones in order to ensure bias-free research to the greatest extent possible. Consensus sessions were used to resolve any differences among the researchers. Involvement of multiple researchers helps to ensure objectivity of data interpretation.

3.3.2 Case Study

Case studies are employed in Article 3, 4 and 5.

In Article 3, an exploratory case study is used to examine how software startups use freelancers for software development. The exploratory research was conducted because the theory about freelancer involvement in startup operations was too limited. The research methodology involves three steps:

- Three software startup founders and senior software engineers were interviewed for this case study. In other words, employees who have recently joined or been linked with the company are not considered for interviews. To find relevant propositions, the three situations were compared (cross-case analysis).

- Case study of 54 freelancers to learn about their experiences working with startups, the challenges they confront (if any), and the perceived value they contribute to software projects. To find significant propositions, the 54 cases were compared (cross-case analysis).
- Comparative analysis of above findings to meet research objectives (cross-case analysis).

Interviews and observations were used to obtain data from the startup instances. The study of archival records was unfeasible due to the fact that the startups (cases) analyzed do not keep documentation records. A flexible interview guide was used to conduct semi-structured interviews. The three startups' working sites were visited for observation.

Questionnaires and group interviews were used to gather information from the freelancers. Initially, the questionnaire was addressed to freelancers (after ensuring ethical implications and gathering their consent for participation). The goal of such a questionnaire is to streamline subsequent levels of online interviews based on the freelancer's initial responses and to create an interview guide accordingly. This is because the freelancers' data was analyzed and classified into groups based on comparable responses. Semi-structured interviews were conducted as a group rather than individually with the groups with similar reports. To elaborate on their comments and clarify any ambiguities, the online interviews were performed using methods such as Skype, WhatsApp, and voice calls. The information gathered was of a qualitative character.

The grounded theory approach was used to analyze the data. The transcripts were coded, and the codes were combined to create a set of propositions/hypotheses and the evidence that supported them. The conclusions were also contrasted to those drawn from qualitative data analysis in other circumstances (cross-case analysis). Evidence that contradicted the propositions aided scholars in revising them.

Article 4 uses an experimental case study to examine how software startups use freelancers to innovate value propositions. Research methodology involves four activities- Pre-study, Data collection, Data analysis and result assessment.

During pre-study, employees from the three startups participated in an online session on "innovation, freelancing, crowdsourcing, and software development" (41 employees of all ranks, including founder). The goal was to inform them about the research study's goals, obtain their consent for voluntary participation, assist in resolving terminological differences between researchers and the startup team, pique employee interest in participating, and finally, establish prior relationships with them that would aid in creating a friendly environment during interviews.

The qualitative data gathered through data gathering was analyzed using the grounded theory approach. By repeatedly constructing and testing hypotheses, the recordings of the interview session were transcribed, coded, and codes merged to generate responses to the formulated study questions.

The findings of this study were given not only to the case study participants (18 employees), but also to a randomly selected group of colleagues who took part in the research's pre-study phase (but not the data collecting) (18 randomly selected employees out of 23 employees). The evaluation of the results is divided into two groups: G1 (18 employees) and G2 (18 employees) (18 randomly selected employees who did not participate in case study except pre-study phase). The evaluation of the results is divided into two groups: G1 (18 employees) and G2 (18 employees) (18 randomly selected employees who did not participate in case study except pre-study phase).

Article 5 featured a case study of startups that have used (or are presently using) freelancers for Requirement Engineering work. Purposive sampling was used in the case study to select startup examples based on their capacity to meet the research study's goals. Data was collected from the startups in India and Switzerland through interviews and observations. Both startup founders and a Product Manager (an experienced developer who has been appointed as a product manager) were interviewed. Field notes and interview sessions were only transcripted at the time of the interview, and grounded theory was used to develop relevant responses to the study questions. These startups were brought in because they had engaged freelancers during the requirement engineering process. The grounded theory approach was used to analyze the data. The transcripts were coded, and the codes were combined to create a set of propositions/hypotheses and the evidence that supported them.

3.3.3   Experience Reports

Experience reports are employed in research study 6. The experience reports are based on firsthand knowledge gained from consultancy engagements, incubator involvements, executive program design, and joint research projects with startups. The research study outcomes are disseminated in Article 9 and 10. Appendix IX contains information on Article 9, and Appendix IX contains information on Article 10.

Research study outcome which is disseminated in Article 9 reports the real consulting experience of the globalisation project of Spanish startup. The experience report reflects the challenges faced in globalisation, strategies executed to find and validate innovative value proposition which meets foreign market needs and implications for product manager are specified to help him innovate in foreign markets. The experience report could be easily adopted or adapted by startup community to drive globalisation based on innovative value proposition. Research study disseminated in Article 10 reports the mutual value that could be generated between startups and academia, driven by practical experiences with academia, corporates, and startups. The experience report is presented as an expert opinion which highlights the business impacts of strategic partnerships between academia and startups.

3.3.4 Multiple Research Methods

Customers can be involved in value proposition innovation tasks through online feedback gathering technologies, according to research study 4, which focuses on how customers can be involved in value proposition innovation tasks. Market investigation utilizing online tools, which is a new activity during a pandemic, is best reported using real-world experiences. To quantify the outcomes, small surveys are also undertaken. Article 6 disseminates the findings (refer to Appendix VI).

The findings of research study 5 were based on an examination of the findings of research studies 1, 2, and 3, as well as short surveys, case studies, and the author's own experiences. Entrepreneurs can use the results to evaluate freelancing platforms. The findings will assist entrepreneurs in strategically selecting freelancers with a long-term focus. Article 7 disseminates the findings (refer to Appendix VII).

Summarising, Article 6 and 7 employs the multiple research methods—Experience Reports, Case study and Surveys. The use of multiple research methods helps to explore the phenomenon under study from multiple perspectives taking advantage of individual strengths of individual research methods. For example, experience reports can be used to report explanatory research findings which could be further investigated using case studies; case studies that are ideal to study phenomenon in its real settings. The gained knowledge can be further validated through surveys with the knowledge experts.

3.4 Unity and Coherence Between Individual Research Study Design

Mixed methods research, i.e., qualitative, and quantitative research, is used in six of the research studies. The research can be classified as multimethod since it uses a variety of research methodologies, such as case studies, questionnaires, experience reports, and systematic literature reviews. The integration of various methodologies aids in triangulation, which allows researchers to gather diverse views on the topic under investigation to improve the reliability and validity of their findings. Individual studies using various research methodologies are designed to explore, describe, and explain the phenomenon under investigation to meet the overall research objectives.

Research study 1 involves systematic mapping study to analyse research trends across value proposition innovation in startup context. As the objective is to identify the research areas that require urgent research efforts and analysis of research types already conducted, the systematic mapping study is more appropriate.

The outcome of this study will drive the need for investigating research trends of freelancer involvement in value proposition innovation activity.

Chapter 2 highlights why this research focuses on freelancers, customers, and strategic partnerships with academia for fostering value proportion innovation in startups. This resulted in employing systematic mapping study in research study 2 for investigating freelancer involvement in value proposition innovation management using the lens of software development activities.

In parallel, the research study 4 reports using practical experiences, for example experience reports, and small surveys conducted with entrepreneurs to identify the online feedback acquisition tools for continuously identifying innovative ideas about value proposition to foster innovations.

The practical experiences with startups during pandemic shape the case study which will be conducted with entrepreneurs followed by the survey to meet research objectives of the study. Outcomes of research study 2 resulted in triggered research study 3.

Research study 3 involved multiple case studies to investigate the "how" and "why" types of questions pertaining to freelancer involvement in fostering value proposition innovation. This also includes the value proposition innovation efforts in global markets as reported in Article 5. The case study was based on the theoretical framework provided by research study 2. The survey was conducted to do member checking to validate case study findings thereby making results more reliable.

The outcome of research study 3 resulted in three Articles 3, 4 and 5. The Outcome disseminated in Article 3 provided a theoretical framework to Article 4. The Outcome of Article 3 and 4 provided a theoretical framework for Article 5.

Research study 5 provides an evaluation framework to evaluate freelancing platforms. The basis of the framework is the conceptualisation of findings of research study 3. The findings are also improved based on case study conducted with entrepreneurs and small surveys to validate the framework.

Research study 6 involves experience reports and reviews to analyse strategic partnerships with academia and customer involvement for fostering value proposition innovation.

Research study 4 provided necessary motivation to investigate customer involvement for value proposition innovation. Further, the role of academia had been reported based on practical experiences in consulting projects.

Overall, the use of systematic mapping studies, case studies, surveys and experience reports resulted in meeting the research objective of the conducted research study.

Figure 3.1 depicts the unity and coherence between individual research studies from the research design point of view.

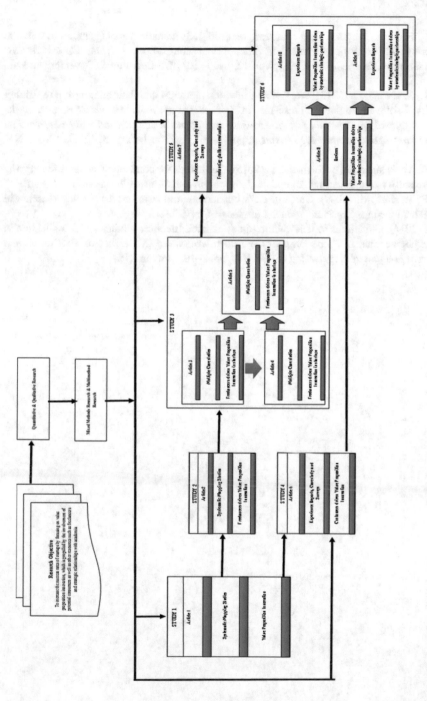

Fig. 3.1 Unity and coherence between individual research study design

References

Petersen, K., Feldt, R., Mujtaba, S., & Mattsson, M. (2008). Systematic mapping studies in software engineering. In: *Proceedings of the 12th International Conference on Evaluation and Assessment in Software Engineering (EASE)*, Vol. 12, pp. 1–10, BCS Learning & Development Ltd., Swindon, UK.

Petersen, K., & Gencel, C. (2013). Worldviews, research methods, and their relationship to validity in empirical software engineering research. In *Proceedings of the 2013 Joint Conference of the 23rd International Workshop on Software Measurement and the 8th International Conference on Software Process and Product Measurement*, pp. 81–89, Ankara, Turkey, IEEE, New York, NY, USA.

Petersen, K., Vakkalanka, S., & Kuzniarz, L. (2015). Guidelines for conducting systematic mapping studies in software engineering: An update. *Information and Software Technology, 64*, 1–18.

Runeson, P., & Höst, M. (2009). Guidelines for conducting and reporting case study research in software engineering. *Empirical Software Engineering, 14*, 131.

Wohlin, C. (2014). Guidelines for snowballing in systematic literature studies and a replication in software engineering. In *Proceedings of the 18th International Conference on Evaluation and Assessment in Software Engineering (EASE '14)*, pp. 1–10, London, UK.

Summary of Appended Articles

4

This chapter summarizes the ten articles that make up the book. These articles provide the results of the research that are the responses to the research questions. These findings have already been published in high-quality international venues, offering a great foundation for disseminating them to the scientific community.

4.1 Relationship Between Research Objectives, Research Questions, and Relevant Research Publications

Table 4.1 shows a graphical depiction of the articles, as well as the research questions addressed, and the research method used.

Figure 4.1 represents the relationship between research objectives, research studies, research questions, and research publications.

4.2 Details of Individual Research Studies and Outcomes

Article 1: Requirement Engineering in Startups: A Systematic Mapping Study.

Purpose: The study's goal is to examine the current level of requirement engineering research in the context of startups, as documented in the literature. The analysis of the research area identifies research trends in order to (a) predict how much support startups can get from the literature in order to improve their success rates, and (b) identify research gaps in order to motivate researchers to conduct future research that could be applied to startup contexts. The focus on analysing the state of the art of requirement engineering is because startups have a high failure rate due to their inability to achieve a suitable

© The Author(s), under exclusive license to Springer Nature Switzerland AG 2022
V. Gupta, *Strategic Value Proposition Innovation Management in Software Startups for Sustained Competitive Advantage*, Synthesis Lectures on Technology, Management, & Entrepreneurship, https://doi.org/10.1007/978-3-031-18322-5_4

Table 4.1 Research results

S. no.	Research study	Research question	Research method	Publication number (appended)
1	Study 1	What is the current state of value proposition innovation in startups?	Systematic literature reviews	Article 1
2	Study 2	What is the current state of freelancer participation in startup value proposition innovation management?		Article 2
3	Study 3	What is the current state of practice in startups for freelancer participation in innovation management activities in terms of methodologies, practices, problems, and real-world outcomes?	Case study	Article 3
				Article 4
				Article 5
4	Study 4	How can value proposition innovation be fostered through customer participation?	Experience reports, case study and surveys	Article 6
5	Study 5	How can entrepreneurs assess freelancing platforms to create and sustain strategic relationships with freelancers?		Article 7
6	Study 6	How might academic strategic partnerships be investigated as a means of promoting innovation management?	Reviews	Article 8
			Experience reports	Article 9
				Article 10

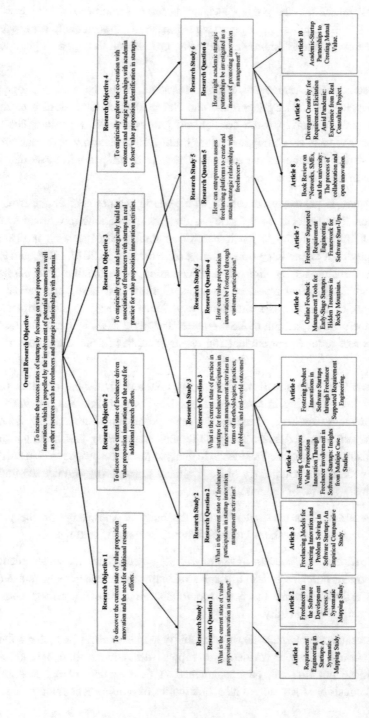

Fig. 4.1 Research objectives, research questions, and relevant research publications

product/market fit. In other words, they provide a solution that falls short of the user's expectations. This means that the product lacks an accurate value proposition capable of increasing client value. Requirement engineering is an activity that can help startup teams in identifying and continuously improving the value proposition that gives high value to users.

Study design/methodology/approach: Studies extracted from the four bibliographic databases (IEEExplore, ACM, Springerlink, and ScienceDirect) and studies extracted using a forward snowballing approach are subjected to systematic mapping. The classification scheme is created by coding individual studies. Information was taken from the abstracts of the studies to populate the schemes that were developed and those that were already available in the literature.

Findings: The study is primarily concerned with generic requirement engineering and product validation. The majority of the study is done as evaluations (empirical studies) with the goal of offering theory to the research community. The research is primarily motivated by the goal of achieving product/market fit. However, in 2017, 2018, and 2019, research studies focusing on requirement documentation, prioritization, and elicitation are losing ground. Studies that report on research solutions that have been validated in laboratory settings or in real-world situations, experience reports, opinion papers, and philosophical papers are all lacking in the literature. The good news is that in the last five years, the number of requirement engineering research studies in the startup setting has increased.

Originality/value: The literature lacks the studies that investigate the research trends across requirement engineering in startup context. The research findings provide future research directions to researchers to improve the value proposition innovation by focusing uniformly across different activities. The requirement engineering had been correlated with value proposition innovation. The idea is to analyse the value proposition in software startups from the technical terminology of Requirement Engineering and use the findings to improve the management of the innovation.

Research implications: This research added to the body of knowledge on the present state of research in value proposition innovation in the context of startups.

Practical implications: The research outcome will help researchers across technical and business management domains to identify areas of innovation management that require urgent research efforts. Further, the need for collaborative research efforts could be valuable for the startup community.

Conclusion: Currently, the literature has limited ability to help startups by giving solutions (for example, research solutions, evidence to support decision-making, best practices, experiences, and so on) that can be implemented in their actual setting. Researchers must maintain a consistent focus across all sub-activities of requirement engineering, with effort distributed across various research types. This helps startups not only by providing validated solutions, but also by providing experience reports, opinions, new conceptual frameworks, and empirical evidence that can help them make better decisions.

Article 2: Freelancers in the Software Development Process: A Systematic Mapping Study.

Context: Freelancers could help speed up the software development process by bringing their specialized expertise to bear on producing high-quality results. They could help businesses (particularly startups) in fostering innovation by suggesting innovative ideas and offering their experience in putting them into action (for instance, designing solutions, coding solutions etc.). Freelancers could function as virtual members of the software development team successfully and efficiently. The organization must make educated decisions regarding which tasks to assign to freelancers, which freelancers to hire, how to price the task, and how to evaluate the work that is submitted. The freelancer, on the other hand, should make an informed decision about determining the monetary value of the task to be charged, trusting the requester, analysing the task's skill requirements (finding matches between skill requirements and skills processed), choosing the best task, and maintaining the highest level of reputation. The literature, on the other hand, does not equip freelancers and businesses with instructions to help them make decisions. Companies will improve their software development metrics like development time, involved costs, if freelancers are carefully picked for the most appropriate work.

Purpose: The purpose of this study is to give research into trends in freelancer-supported software development to the academic community. This aids in determining which software development domains have a higher concentration of research activities, which areas have empirical data to support management decision-making, and which areas require additional research.

Study design/methodology/approach: Planning the mapping methodology, implementing the protocol, and reporting the findings using various visualization tools such as bar charts and pie charts are all part of the systematic investigation. On four bibliographic databases, the search procedure was expected to be carried out using a set of inclusion and exclusion criteria (IEEExplore, Springerlink, Sciencedirect, and ACM digital library). The papers that are relevant are chosen using inclusion and exclusion criteria. To incorporate the additional relevant papers, the google citations of relevant papers are subjected to inclusion and exclusion constraints once more. After examining the abstracts of the investigations, the systematic schema was built and populated.

Findings: The findings reveal the following:

(a) Rather than individual life cycle activities, the research focuses on generic software development (78%).
(b) There are a finite number of empirical studies (25%).

(c) The literature lacks a number of studies presenting solutions and testing them on real-world instances in industrial settings. In comparison, 72% of validation procedures (i.e., solutions evaluated in lab settings) were used.

(d) At the moment, the literature has limited ability to provide advice (in the form of opinions and experience reports) for incorporating freelancers in the software development process to software organizations (including startups).

(e) Collaboration and Coordination (33%), Developer Recommendation (or selection) (19%), Team Formulation (14%), Task Recommendation (allocation) (14%), Task Decomposition (11%), Privacy and Security (Confidentiality) (11%), Budget Estimation (8%), Recognition (8%), Trust Issues (8%), Market Dynamism (6%), Intellectual Property Issues (6%), Participation (6%), (3%) are reported challenges in literature. These challenges are very interactive, and each one influences the others.

(f) The researchers' most recent focus (a total of 7 studies in 2019) has been on generic software development, which includes collaboration and coordination (3 studies out of 7), developer recommendation (2 studies out of 7), and task suggestion (2 studies out of 7).

Originality/value: The literature lacks the studies that investigate how freelancers could be employed across value proposition innovation activities. The value proposition innovation has been seen under the lens of the software development process. The outcome provides directions for future research that could improve freelancer integration across entire value proposition innovation thereby helping startups to overcome their resource limitations in a cost effective manner and gain market success.

Further, the correlation between software engineering and value proposition innovation management is not explicitly defined in literature. The outcome provides a new way of looking at value proposition innovation by learning lessons from technical terminology as prevalent in software startups/companies. The outcome will improve the Gig economy and startup success rates.

Research implications: This study contributed to the corpus of knowledge on the current state of value proposition innovation research in startups.

Practical implications: The findings of the study will aid researchers in identifying research topics that require immediate attention. By bridging the research gaps, they will be able to identify creative ways of outsourcing innovation management activities to freelancers.

Conclusion: Researchers are interested in the freelancer-driven software engineering study subject, but it will take a long time for it to mature. More empirical studies and evaluation-based solution research are urgently needed to help firms (especially startups) in fostering innovation. Furthermore, to address the unique issues associated with individual activities, the research effort should be evenly dispersed across the various development phases. The effective management of freelancers in software development could help businesses and startups in fostering innovation and remaining competitive in the marketplace.

Article 4: Freelancing Models for Fostering Innovation and Problem Solving in Software Startups: An Empirical Comparative Study.

Context: By increasing software development criteria such as cost, time, and quality, freelancers and startups could supply one other with exciting prospects that lead to mutual progress. Freelancers with specialized talents could help startups in reducing the risks associated with new innovations and markets, as well as the ability to respond rapidly to market concerns (and with higher quality). This necessitates long-term, trust-based relationships between freelancers and startups, as well as promising agreements based on motivation (leading to the growth of both parties). If freelancers are chosen using data-driven decision-making, they can help firms stimulate innovation and complete software development projects more efficiently than in-house teams.

Purpose: The paper's three goals are to (1) investigate startup strategies for outsourcing software development tasks to freelancers (dubbed "freelancing association strategies"), (2) identify challenges in such outsourcings, and (3) determine the effects of outsourcing tasks to freelancers on overall project metrics. The overarching goal is to learn about ways for incorporating freelancers in the software development process throughout the startup lifecycle, as well as the challenges and benefits that come with it (to maintain competitive advantages).

Study design/methodology/approach: This paper uses case studies of three software firms in Italy, France, and India, as well as a survey of 54 freelancers, to conduct empirical research. In order to identify association models, concerns, obstacles, and reported findings originating from such associations, the data is studied and compared. Members double-check the case study outcomes with the research participants, resulting in a higher level of result agreement.

Findings: The findings suggest that the freelancer association approach is task-based, panel-based, or a combination of the two. Issues such as establishing pricing, setting deadlines, difficulty in finding good freelancers, quality issues with software artefacts, and efforts to gain access to freelancer job submissions for compensation confine the associations. If good freelancers are available, associations have a favourable impact on software development (which lasts long for various tasks). Finally, the research presents a freelancing model framework and suggests activities that could improve the situation for both sides and streamline such relationships. Fostering innovation in startups is thus a trade-off situation that is hampered and aided by a slew of competing factors.

Originality/value: The empirical studies of freelancer involvement in innovation management with startups are missing from literature. The outcome will help the startup community to develop better strategies to associate with freelancers by overcoming the challenges as observed by their peers.

Research implications: This study added to the body of knowledge on freelancer partnerships with startups to help them in market innovation.

Practical implications: The study's findings will help entrepreneurs overcome resource constraints by involving freelancers in the design, implementation, and commercialization of their product offerings.

Conclusion: According to the findings of the report, entrepreneurs use a variety of techniques to outsource innovation activities to freelancers, but they encounter greater hurdles in doing so. Due to its stochastic nature, resource constraints, and newness in the industry, startups have trouble managing the freelancing process. Association with a good freelancer from the start of a project has an impact on subsequent long-term relationships. Long-term relationships are beneficial to the growth of a startup. Informed judgments on freelancing outsourcing, such as the jobs to be outsourced, perceived pricing, duration, and so on, will encourage freelancers to participate actively. Because this is a two-sided market, both entrepreneurs and freelancers must offer unique value propositions to one other in order to build and maintain long-term relationships.

Article 4: Fostering Continuous Value Proposition Innovation Through Freelancer involvement in Software Startups: Insights from Multiple Case Studies.

Context: By integrating freelancers as a source of creative ideas (that boost customer perceived value) and as specialists for implementing innovative ideas, software firms could continuously innovate their business model value proposition (by undertaking software engineering tasks). Startups use one of three ways to work with freelancers: task-based (association stops when the outsourced task is completed), panel-based (task is outsourced to a panel of freelancers linked with the business), or hybrid. The main inhibitors for associations with freelancers are uncertainty, terminology issues, high technical debt, lack of documentation, lack of systematic decision-making processes, lack of resources, lack of brand values, and the need for the freelancer's continuous involvement to incorporate continuous validated learnings, merging freelancer perspectives, and deciding the level of their involvement in individual requirement engineering (or value proposition innovation) activities. Good freelancers' availability, as well as their long-term and consistent commitments, are essential for value proposition innovation. The theory regarding freelancers and software companies is expanded by looking at real-world examples of startups that have effectively used freelancers for value proposition innovation by capturing innovative ideas and gaining the freelancer's abilities to put those ideas into action.

Purpose: The purpose of this article is to describe the tactics used by software startups to encourage value proposition innovation by incorporating freelancers on a regular basis, as well as how they deal with the issues that arise as a result of the associations. The conclusions are based on a review of real-world practices of businesses that have shown to be successful in the market by utilizing freelancers and implementing ongoing innovations that have resulted in increased market share.

Study design/methodology/approach: This paper conducts empirical research using case studies of three software startups in Italy, France, and India that are on the verge of becoming giant enterprises with significant market share. The present practices are reported, demonstrating the successful approach of executing freelance association strategies for value proposition innovation as well as how to manage the problems that arise. The findings are compared to those of two young businesses situated in Switzerland and India, in order to provide helpful lessons for the young companies. Employees from the startups under investigation confirm the findings of the case study (both those who participated in data collection and those who did not).

Findings: The findings show that including freelancers in value proposition operations, which is a core company activity, benefits both freelancers and entrepreneurs. The investigated startups were able to overcome generic and value proposition specific challenges associated with freelancer associations. By properly employing freelancers evenly throughout value proposition activities, startup teams could cut development costs, shorten time to market, and boost customer satisfaction (by supplying features that match real market demands). If done in collaboration with freelancers, entrepreneurs could manage innovation with small teams (relative to human resources in firms), allowing team members to learn new talents, improve their skills, and gain new insights into their industries. When the level of freelancer involvement across numerous value proposition activities is higher than when they are simply involved in a few activities, the business consequences are bigger. Although the researched businesses are not entirely reliant on freelancers, their perspectives and abilities are considered as a valuable source of market success. Involvement of freelancers is seen as a chance to have access to global market perspectives that would otherwise be difficult for in-house teams to obtain. They also help marketers reach customers by resolving technological debt, improving skills for future innovation, and resolving technical debt (promoting product and gaining access to the feedback). Overall, the studied start up's value proposition innovation has varying levels of freelancer involvement, but these startups have reported positive impacts on the business in terms of development cost reductions, shortened time to market, and high customer satisfaction as evident through early product/market fit and rapid market growth achieved by the startups. Employees at the startup confirm the findings of the case study (member checking). Box plots are used to analyse the replies obtained, revealing a higher level of result agreement among the personnel.

Originality/value: The empirical studies of freelancer involvement in startup value proposition innovation management are missing from literature. The outcome reports the successful strategies of startups to involve freelancers in value proposition innovation and the ways to overcome the associated challenges with such associations. The outcome will become lessons for their peers, helping them to innovate the reported practices and adopt them to their unique working context.

Research implications: This study added to the body of knowledge on freelancer partnerships with startups to help them in launching innovative product or service in the market.

Practical implications: The study's findings will help entrepreneurs overcome resource constraints by involving freelancers in the design, implementation, and commercialization of the innovative value proposition of the product or service.

Conclusion: The paper's findings show that by utilizing freelancers as a source of creative ideas and as talent resources for putting those ideas into action, firms may successfully develop their value propositions and sustain competitive advantages. Involvement in the value offer necessitates a long-term commitment from freelancers with the flexibility to increase their learning over time, culminating in market success. This necessitates the startups developing trusting relationships with them and motivating them to participate on a regular basis. Startups may prefer not to rely entirely on freelancer assistance, but rather to supplement their own knowledge with the viewpoints provided by freelancers. The assistance of freelancers during implementation may help startups in implementing creative ideas optimally, supporting skill upgrades of the in-house team, promoting teamwork, fostering trust, and ensuring that high coding standards are adhered to. The business impact of freelancer involvement is determined by the level of involvement evenly throughout numerous value proposition innovation activities, not by the level of involvement across a few activities.

Article 5: Fostering Product Innovations in Software Startups through Freelancer Supported Requirement Engineering.

Context: In highly competitive markets, encouraging value proposition innovation (or evolutionary requirement engineering) is the key to gaining a competitive advantage. Freelancers could provide unique product value proposition ideas as well as skills in putting these ideas into action.

Purpose: This research paper investigates freelancers' involvement in requirement engineering activities in order to innovate value propositions continuously and apply their knowledge to diverse requirement engineering tasks.

Study design/methodology/approach: This paper presents the findings of a case study conducted with companies that use freelancers for requirement engineering. The data are then compared to the literature to have a better understanding of the freelancer-assisted requirement engineering sector.

Findings: According to the findings, freelancers can assist in the innovation of value propositions by bringing varied perspectives on global segments as well as skill in performing requirement engineering tasks. Depending on the startup context, freelancers have variable amounts of involvement in requirement engineering activities and are heavily challenged by various impediments. The difficulties in selecting the best freelancers, assuring their long-term association for ongoing rework deriving from market learnings, creating trust, mechanisms to integrate their perspectives, establishing communication, negotiations, and strategic pricing are among the barriers.

Originality/value: There are no studies involving freelancers in value proposition innovation in global marketplaces. This study expanded on the knowledge shared in the previous two studies (articles 4 and 5), but in the context of global marketplaces. By employing freelancers in the manner described in the article, the outcome will help the startup community in taking advantage of globalisation.

Research implications: This study added to the body of knowledge on freelancer partnerships with startups to help them in market innovation in global markets.

Practical implications: The study's findings will help entrepreneurs overcome resource constraints and globalise their business operations by involving freelancers in the design, implementation, and commercialization of their product offerings that can satisfy foreign market demands.

Conclusion: Due to resource constraints, it is necessary to build freelancer involvement from the start of the startup life cycle with a promise of long optimally-term rewards in exchange for their trustworthy and correct viewpoints, which is more difficult to obtain by enlisting crowds of customers. More research is needed to see how freelancers could represent samples of worldwide scattered customer segments as an input source of information on the one hand, and become startup team representatives on the other, to help startup expand its operations in foreign markets.

Article 6: Online Feedback Management Tools for Early-Stage Startups: Hidden Treasures in Rocky Mountains.

Context: Value proposition innovation is the key for competitive advantage and is a continuous activity. The early-stage start-ups do perform continuous experimentations to validate their assumptions about the business model; value proposition being the major element, by having continuous face-to-face interactions with the customers in co-located physical space. The feedback is crucial for reaching product/market fit and driving the incremental software development. Face-to-face interactions in the same physical space being limited during pandemic, forced start-ups to use online tools for feedback acquisition to do release planning based on their analysis.

Purpose: This article explores the area of feedback management tools in the context of early-stage start-ups by providing insights from real experiences about these start-ups during the pandemic; the findings that could set the stage for the "new normal" and adding pertinent information with much industrial relevance to the literature.

Study design/methodology/approach: This study is premised on the authors' practical experiences and the insights offered by the start-up team on how to manage feedback during a pandemic, both quantitatively and qualitatively. This is made possible by the authors' ongoing connections with a variety of early-stage European startups in the form of consulting projects, incubator participation, executive program creation, and cooperative research initiatives with Startups as partners.

Findings: The findings show that, to include remote customers in a customer development process, feedback gathering technologies should assist both customers and startups in comprehending communicated product information and making useful suggestions that could encourage innovation. Integration, proximity, social interactions, individual targeting, analytics assistance, platform integration, and multiple functionalities across platforms are the major utility characteristics that tools should deliver. Among the several options, the one that gives simplicity of use, utility, and low investment should be used in development policies.

Originality/value: The research focuses on the use of technology to connect customers with startup teams for knowledge acquisition during pandemics. The studies about this topic are almost negligible in literature. The end result will assist the startup community in co-creating with clients through the use of proper feedback acquisition technology.

Research implications: This study added to the body of knowledge on online tools that could integrate customers in market research activity of startups for continuously innovating their product.

Practical implications: The study's findings will help entrepreneurs to limit their co-located meetings and focus more on online interactions with customers to evolve their product offerings in the market.

Conclusion: There is untapped potential for companies to make greater use of available tools for feedback management, such as gathering feedback, automating analytics, and more, resulting in real data-driven business decisions. We also expect individuals who take advantage of it to do better in the future, extending the lessons learned during the pandemic to the new normal.

Article 7: Freelancer Supported Requirement Engineering Framework for Software Start-Ups.

This article provides the framework for conducting value proposition innovation or requirement engineering activity with the help of freelancers. This article divides the pre-outsourcing, outsourcing, and post-outsourcing phases of freelancer support Requirement Engineering (RE) into three parts. Each phase consists of several actions, which are either carried out directly by the startup team or using freelancing platforms' services. These actions, in combination with other criteria such as simplicity of use and utility, form the comprehensive evaluation framework for selecting freelancing platforms, freelancers, and job deliverables for RE. Financial ratio analysis is used to validate the methodology based on the business consequences of outsourcing and post-outsourcing phases in Spanish startup environments.

The outcome demonstrates the framework's high applicability in improving freelancer-driven Requirement Engineering, which leads to a healthy business.

The study of freelancer-supported RE activity has the potential to change the fate of the startup community, which is nonetheless plagued by high burnout rates. The more integrated set of functions and services supplied by freelancing platforms maybe able to help companies in leveraging the knowledge spread across global freelancing professionals, who can serve as excellent customer support representatives as well as RE experts.

The Freelancer supported RE should be considered holistically by freelancing platforms, which should provide services for many tasks associated with the process. The set of services given serves as an evaluation framework, assisting the company in determining which platforms to use depending on their business goals and internal capabilities. The framework was developed using real-world startup methods and tested in Spanish startup environments. Due to the unique working environment of startups, the framework requires further validations to provide relevant insights based on generalized business outcomes throughout the startup community, but in its current state, it is very adaptable and suitable for serving the needs of startups.

Future research problems unique to these phases, as well as freelancing platforms, are identified, along with the need for more multidisciplinary research.

Article 8: Book Review on "Networks, SMEs, and the university: The process of collaboration and open innovation.

The study entails a review of a book about startup engagement with academia in the United Kingdom. The findings suggest that cooperation between academia and entrepreneurs in the United Kingdom are an excellent example of effective collaborations. These kinds of coalitions could be formed and maintained in a systematic way.

Article 9: Divergent Creativity for Requirement Elicitation Amid Pandemic: Experience from Real Consulting Project.

Context: Pandemics influence companies' ability to find product features that match market needs; this is an activity that necessitates close engagement with clients in the same physical place. Although online tools can help overcome this barrier, early-stage firms often have insufficient resources and lack access to potential clients, limiting their online interactions. To identify Requirement Elicitation methodologies and technologies that could help startups identify product/market fit with minimal same physical space engagement with consumers, divergent creativity is required. Open innovation involving universities, specialists, and researchers maybe able to assist companies in meeting market demands.

Purpose: This paper describes one of the author's consulting experiences with a Madrid (Spain)-based startup that successfully identified its market during a pandemic using market research based on secondary studies, primary research involving potential clients (or users) via online means, and limited interactions in the same physical space.

Study design/methodology/approach: This paper details the author's experience with a firm established in Madrid (Spain) that tested its technology-based product in a foreign market.

Findings: Daily brainstorming sessions with a group of researchers, specialists, and academics resulted in a variety of divergent ideas for identifying markets in the midst of a pandemic and putting them to the test in a real-world setting, all of which proved to be beneficial for the startup to globalise its business operations beyond Spain.

Originality/value: In the literature, there are negligible studies on performing market research to develop and validate value propositions in global marketplaces. With secondary research and a little primary research, the outcome will help the startup community in globalizing their operations, designing, and validating their value proposition for global markets. Overall, a sensible globalisation decision with novel value proposition will be made in an efficient manner.

Research implications: This study contributed to the body of knowledge on how secondary research, when combined with primary research, can help startups in exploring overseas markets and developing products for those markets. Furthermore, this teaches us how academia, freelancers, professionals, and customers can all play a role in globalisation operations.

Practical implications: The conclusions of the study will help entrepreneurs in having open innovation in order to globalize their corporate operations.

Conclusion: Due to social distancing norms, need elicitation; the source of creative ideas for the product; is difficult to conduct during the pandemic. However, startup-academia collaborations may provide startups with simple access to client needs that might otherwise be difficult to fulfil by an in-house staff. Working in groups and having regular brainstorming sessions helps to produce a wide range of unique ideas and converge them into significant insights for the firm. These gatherings were crucial in bridging knowledge gaps and disseminating market information. The use of a combination of market research approaches that included both online and offline contacts with customers, as well as secondary market research and expert advice, greatly aided in the development of market understanding. In the face of a pandemic, interactions with potential consumers must take place in a more uncertain environment, thus it's more vital to improve your market understanding by asking "right questions" to "right customers" rather than focusing on complex tool selection. The team must be adaptable when it comes to tool selection, which is heavily influenced by the context of the consumer. Evaluation of demand in a new market for current products necessitates face-to-face interactions with potential customers to determine their needs for possible items, followed by adaptation of the offering to the new market (also called as customization of the product). The support of academia could be a game changer for resource-strapped startups, particularly in terms of providing a roadmap for market research, generating, and implementing creative ideas, and overcoming customers' apprehensions about participating in Requirement elicitation activities due to the early stage startups lack of branding due to their strong academic reputation.

Article 10: Academic-Startup Partnerships to Creating Mutual Value.

The foundations of the innovation ecosystem are academia and startups. By collaborating as strategic partners, academia and startup businesses may achieve reciprocal, long-term benefits through two-way knowledge transfer. We suggest that this may be accomplished without the involvement of huge corporations, which could stymie the missions of universities and startups, impeding innovation growth. Startups and universities might work together to facilitate knowledge transfer in a more flexible, agile, and less bureaucratic fashion, allowing both parties to benefit from strategic relationships.

Results and Contribution

This chapter introduces the outcomes of the research. The outcomes of individual or multiple research studies for each research question are presented. Finally, the discussion section presents the in depth exploration of the research outcomes. The in depth exploration aims to highlight that how freelancers could be integrated as virtual team members of startups to foster value proposition innovation strategically. Further, how such partnership with freelancers and academics could be executed for long run and it could be a tool for competitive advantage. In other words, how the freelancers and academia without any legal long term relationship could be employed for sustainable competitive advantage in highly competitive marketplaces. Finally, the implications of the conducted research for entrepreneurs, customers, freelancers, academia, and the government are discussed.

5.1 Result Analysis

The outcomes of individual research studies which answers the formulated research questions to meet research objectives are briefly discussed below:

RQ1. What is the current state of value proposition innovation in startups?
The value proposition innovation help startups to improve the customer value either through improved existing product or as a new product or services. This term is closely aligned with "Requirement Engineering" terminology as used in software engineering. The software startups deliver customer value through their software solutions as products or services. The value proposition innovation involved identifying the new ideas about enhancing product value proposition, implementing these ideas using suitable software development process models and finally commercialising the solutions in the market.

© The Author(s), under exclusive license to Springer Nature Switzerland AG 2022   67
V. Gupta, *Strategic Value Proposition Innovation Management in Software Startups for Sustained Competitive Advantage*, Synthesis Lectures on Technology, Management, & Entrepreneurship, https://doi.org/10.1007/978-3-031-18322-5_5

There are many ways of enhancing the customer value or in other words, performing value proposition innovation. This includes the following:

- Startups could innovate their process models, also called as process innovation. The process innovation could result in better quality product, for instance, higher reliability, lower costs, better time to markets, providing value to startups. This value can then be passed to customers resulting into enhanced customer value, for instance as reduced product prices, higher quality product and much more.
- Startups could innovate their products by launching new products to address new needs or improving existing products by making customer interaction with it easier or meeting their continuously changing needs. In either case, the product innovation aims to identify customer needs and finally identifying the solution to address them to provide them the benefits expected by the customers.

The value proposition defined why the customer will buy or will continue to use the startup product. Continuously innovating the value proposition will be the key to sustained competitor advantage. In some cases, value proposition innovation may not result in product innovation. In management literature, it is reported that the limited research efforts are invested in business model innovation (Ibarra et al., 2020; Pucihar et al., 2019). The value proposition innovation, a part of business model innovation is still an unexplored area. Requirement Engineering which aims to identify the new customer requirements for existing or new product is the key activity to innovate value proposition. The way software startups conducted this activity for value proposition was an unexplored issue. The idea to learn the ways value proposition innovation is reported in management and engineering literature could provide good empirical knowledge to startup community that are finding hard to survive in highly volatile markets.

The outcome of the systematic mapping study conducted across four bibliographic databases, for instance, IEEExplore, ACM, Springerlink and ScienceDirect as well as the research studies identified through forward snowballing, can be conceptualised across two main points namely:

- **Empirical literature support for startups in terms knowledge transfer to improve their success rates**: The literature at this moment has limited ability to provide sufficient value proposition innovation related knowledge to startups. For instance, literature has limited research solutions, empirical evidence, best practices, experiences, that could be adopted by them to improve their value proposition innovation practices. The requirement elicitation activity is synonym to value proposition idea elicitation, requirement prioritization is like ranking the value proposition innovation ideas, requirement validation is like validating with customers the identified value proposition ideas and requirement documentation is documentation of innovation ideas that could be readily used for implementations. For value proposition innovation, the theory underpinning all these activities is important for successfully implement and commercialise the most promising ideas to generate good customer value leading to improved business financial strength in the market. However, the researcher efforts invested in this activity is not uniformly distributed across all these sub activities. Further, the type of research efforts should provide innovation knowledge from different perspectives, for instance, validated solutions, experience reports, opinions, new conceptual frameworks, and empirical evidence. This is because, startups have different working contexts, and they not only require large theoretical knowledge bases but also different types of knowledge to be able to adopt them as per the context unique to them.

- **Research gaps in literature that require urgent research efforts**: The research is mostly focused on generic requirement engineering and product validation activities. The research is conducted mostly as evaluations, for instance empirical studies, with the outcome of providing theory to the research community. Major underlying motivation of the research is to attain the product/market fit. However, research studies focusing on requirement documentation, prioritization and elicitation are losing focus from 2017, 2018 and 2019, respectively. Identifying the innovative ways of capturing innovation ideas from customers is losing researcher focus. Big companies may have access to plenty of ideas, for instance, using internal crowdsourcing or through their access to the extended network of suppliers, existing customers etc. However, in startups, the resources like financial, human resource, brand power are limited, and it may be possible that there is no existing market yet for the proposed product. The ability of these startups to identify innovative value proposition seems to be less supported by the existing body of literature. The literature lacks the studies that report research solutions which are validated in laboratory settings or in real contexts, experience reports, opinion papers and philosophical papers. The positive side of the finding is that the number of requirement engineering research studies in a startup context have increased in the past five years but not matured enough to empirically support startup community.

The outcome of this research indicates that there is a need to find innovative ways of doing value proposition innovation that works in startup context. This includes better ways to identifying new value proposition ideas, their implementation and commercialisation.

RQ2. What is the current state of freelancer participation in startup value proposition innovation management?

The freelancers could be a great resource for the resource stripped startups in identifying innovative value proposition as well as its implementation. They are the key to increase dynamic capabilities of startups to seize opportunities in business environment. The gig economy is on rise especially during pandemic, providing ability to startups to innovate their business model as survival strategy. There is rising number of professionals with rich technical competencies and expertise in application domain that are opting for freelancer work. This had provided opportunity for rise in businesses providing services through digital platforms that connect startups with these freelancers. Amid the rising number of freelancers and the freelancing platforms, the outsourcing decision is rather complex. The startups find it harder not only to identify which activity to outsource but also to whom to outsource and how to turn this outsourcing into strategic long term association. This is answered by Research question 3 and 4 using multiple case studies.

Further, it should be investigated that how much support these startups can get from the body of knowledge. For this to happen, it is necessary to systematically search the bibliographic literature to analyse the research efforts pertaining to freelancer involvement in innovation activities, in other words, value proposition identification and implementation. This is answered by research question 2 using systematic mapping studies. The objective is to investigate the research trends in freelancer involvement in value proposition innovation in software startup context. This will highlight the area of value proposition identification and their implementation using the lens of generic software development activities.

The systematic mapping was conducted over four bibliographic databases, IEEExplore, Springerlink, Sciencedirect, and ACM digital library as well as articles identified through google citation analysis.

The empirical study indicated that literature has research efforts concentrated on freelancer involvement in generic software development rather on individual life cycle activities, for instance, value proposition identification. This indicates that startup community relying on empirical findings to find innovative ways to find innovative value proposition in pandemic or beyond, are likely to be less benefitted. Further, the empirical studies including case studies or experience reports drawn from experts or from industries are too limited. This means that startups will find it hard to find an online platform to identify the best practices reported by their peers or big companies which could help them to adopt them to solve the problem unique to their context. Adding to this issue, the body of knowledge has negligible studies that provides industrially validated solutions to startup community. This signify that startup looking to innovate their value propositions based on freelancer involvement will find almost no support from literature in terms of getting access to new solutions that were valuable to other businesses. The availability of laboratory validated solutions could be meaningful but not large enough to provide much greater support. Further laboratory validated solutions could be seen an indication to follow rather a path already tested in real settings by other businesses.

In literature, the reported studies indicates that startups that collaborated with freelancers faced several challenges like Collaboration and Coordination, Developer selection, formulation of team, allocation of tasks to be outsourced, decomposing tasks into smaller tasks, Confidentiality, estimating outsourcing budget, Recognition, establishing trust, dynamism of markets, Intellectual Property Right Issues, crowd worker involvement, and outsourcing platform capacity utilization. These challenges are highly interactive, and each challenge impacts all other challenges. These challenges will hinder the strategic outsourcing of value proposition innovation tasks to freelancer community and build long-lasting relationships with them.

The overall outcome of the empirical study is that to involve freelancers in value proposition innovation, it is required to speed up the research in this direction in form of empirical studies and solutions validated in real settings. These studies will help startup community to gain knowledge from their peers leading to their market success. Further, it is required for the research efforts to be uniformly distributed across value proposition innovation activities rather providing abstract understanding of overall innovation.

RQ3. What is the current state of practice in startups for freelancer participation in innovation management activities in terms of methodologies, practices, problems, and real-world outcomes?

The research addressing RQ1 help to establish the ground for more research efforts in value proposition innovation based on limited support provided by body of knowledge. Research outcome addressing RQ2 help to provide grounds for more research efforts in direction of strategic freelancer involvement for value proposition innovation. The outcome also indicated that it is important to disseminate the insights about freelancer support value proposition innovation as practised by startups and the associated challenges incurred and benefits reported. This could provide good empirical lessons for the startup community, helping them find grounds to adopt the practices of their peers.

This provided directions for investigating that how freelancers could be successfully integrated with startup teams to drive value proposition innovation. The involvement could help startup to avoid the burnout risks and build dynamic capabilities to turn market opportunities into their business success. This research direction is pursued through multiple case studies conducted with startups located in Europe and Asia at different time intervals.

First empirical research study involves multiple case studies with three software startups located in Italy, France, and India, followed by a survey of 54 freelancers. The objective is to investigate the research topic from a generic context, in other words, not specific to value proposition but to generic innovation management in software startups. This involve investigating how startups associated with freelancers, what association issues as well as challenges are reported and how does this association is valuable for their business.

The outcome indicate that startups executed either of the three association strategic with freelancers i.e., task based, panel based, and hybrid. The panel based and hybrid strategy suggest that in startup context, it is possible to maintain the pool of freelancers with long term focus so that outsourcing could be done among the pool members. This however, can be made more valuable if cross compared with the outsourcing quotes from the freelancing platforms.

These associations require startups to be aware of issues like estimating outsourcing pricings, time estimations, accessing good freelancers, outsourcing work quality, and efforts to access freelancer work submissions for reward. These associations if carried out correctly can have positive impact on business. This however require both parties to develop and maintain long-term relationships. The association should be seen as a strategic process rather than a onetime need-based affair.

The results were based on the practices undertaken by the studied startups and were being continuously innovated by them. The innovation involves continuous process of implementing, measuring results, learning from results, and incorporating validated learning to further improve practices until the freelancer driven value proposition innovation is adjusted enough to bring success for these startups in the market. In the period of time when this case study was conducted, these startups which were in the mid of the growth phase of the startup life cycle. After the series of continuous experimentations, the practices were innovated to better levels that were more matured and tested in real markets. A new case study was conducted with these startups as well as involving two young startups based in Switzerland and India to extend the previous knowledge about freelancer involvement in innovations. This new empirical study focuses specially on value proposition innovation driven by freelancer associations.

The objective was to investigate the association strategies further, challenges and business impacts that could have evolved with the time when previous case study was conducted for value proposition innovation. These findings will be more reliable as the three software startups located in Italy, France, and India that participated in the case study, were at verge of turning into bi companies with great market share. Their market success is an indication of the validation of the freelancer driven value proposition innovation in real settings.

The results indicate the ability of startups to do value proposition innovation through successful integration of freelancers in the business activities. The value proposition activities that had witnessed successful integration of freelancers include-problem exploration, problem prioritization, Minimum Viable product development, documentation, Prototype development, Model designing and coding. These activities of value proposition innovation are core business activities, yet freelancer integration had been the reason with startups to build dynamic capabilities and gain market success through reduced development costs, reduced market launch timings and enhanced customer satisfaction based on product/market fit.

The successful integration had worked because the startups took it as an opportunity by building their internal competencies by learning new skills, improving existing skills, and gaining new market perspectives ways of looking at the potential market from the freelancers. The value proposition innovation has several activities which should involve freelancers rather only those where startups lack the skills in-house. The freelancer association had been a reason for great commercialisation for the products as they helped startups in promoting their products in their professional and social network thereby helping them build a brand image.

The idea of freelancer driven value proposition innovation can go one step higher. The startups can take advantage of not only the rich competencies of freelancers but also their global diversity. The global diversity of freelancers can help startups to globalise their business operations by getting the access to foreign market information.

In another empirical study conducted using case study method with startups that employ freelancers for value proposition innovation activity. The findings are then compared with the already conducted empirical studies, study 2, 3, 4 and 5. The results indicate that the freelancers can help startups in value proposition activities for both global and domestic markets. Their involvement helps startups to gain access to foreign market information which help them to get new ideas, validation support and implementation. These startups have involved freelancers in different value proposition innovation activities but is challenged by their ability to optimally select freelancers, achieving long term associations, maintaining, and enhancing trust levels, the integration of freelancer perspectives with those hold by startup team, communication with freelancers followed by negotiations and outsourcing pricing.

The long term association is an important factor not because it helps to build internal competencies but also because initial startup activities involve series of experimentations that need continuous rework, for instance, evolving prototypes and associated product software code, that require continuous involvement of same freelancers. The successful business outcomes depend on the involvement of freelancers from the beginning of the startup life cycle. The freelancer's global diversity helps them to be source of market information which is local to them but global to startups as well as startup team representatives to establish direct interactions with global customer segments.

The three empirical studies indicate the promising role of freelancers to build capabilities of startups in fostering value proposition innovation. However, the real business impact of such partnerships is evident only through long term and sustainable partnerships.

RQ4. How can value proposition innovation be fostered through customer participation?

The startups launch the software product that best matches the customer needs. To have this fit between the product and the market, they should involve customers in identifying and innovating the product value proposition. The well-known theory of customer development advocates direct communication with customers to better understand the market needs to deliver the product meeting those needs. However, leaving the startup premises and going out in the market to interact with the customers is harder during pandemic due to social distancing and lockdowns. In startups, the customer involvement in product innovation and/or business model innovation usually relies on feedback derived from face to face onsite interactions. Such interactions are hindered by pandemic imposed restrictions.

The pandemic arrival was unforeseen which did not give startup community enough time to make a quicker and rational response. As the time progressed, the different startups innovated the way they involved customers in value proposition innovation activities going from traditional onsite meetings to completely virtual ones. Through consultancy projects, incubator involvements, executive program design, and cooperative research initiatives with Startups as partners, this research study collected qualitative and quantitative perspectives. These perspectives were analysed to identify how startups employed the feedback acquisition tools to connect with remote customers during pandemic.

The research identified the feedback channels that worked for these Startups, their perspectives about the value expected from online tools for feedback management (utility factors), start-up team experiences with online tools as represented quantitatively against utility factors and useful lessons learned from adoption of online tools, which provide hope for the new normal. The findings show that, to include remote customers in a customer development process, feedback gathering technologies should assist both customers and startups in comprehending communicated product information and making useful suggestions that could encourage innovation.

Feedback acquisition technology can be adopted using four selection criterias namely, ease of use for both startups and customers, usefulness, and involved investment. The functional utility factors that startups expect from the feedback tools to provide includes, Integration, proximities, social interactions, individual targeting, analytics support, integration across platforms, and multiple functionalities across platforms.

The most widely employed feedback acquisition tools during pandemic by the startups include social networking sites like Facebook, Twitter, LinkedIn, WhatsApp, Email, Websites, for instance developed using wix.com and wordpress.com, Slack, Google forms, Skype and Zoom. Based on feedback provided by 89 European Startups about the feedback tool ratings as per the functional utility factors is shown in Table 5.1. Figure 5.1 represents the evaluation of feedback acquisition tools against the selection criterias.

The following are observations from Table 5.1 and Fig. 5.1:

- Comparative analysis of the feedback acquisition tools against Functional utility factors indicate that social networking sites ranks better across all factors compared with other tools (Table 5.1). This is because these tools have network effect i.e., they generate more value when more users join the network. This strengthens social proximities making it possible for startups to expand their branding across social networks of their customers. Further, they have inbuilt support for analytics as well as different ways of interacting with the potential customers.

  However, as they are services provided by third parties, it is harder to modify the underlying software code to link multiple platforms together. The stronger social interactions across these sites will be an opportunity with startups to foster interactions with customers through prototyping.

Table 5.1 Comparative analysis of online feedback acquisition tools (© 2021 IEEE, republished with permission from the publisher)

Tool	Utility factor						
	Integration	Proximities	Social interactions	Individual targeting	Analytics support	Integration across platforms[a]	Multiple functionalities across platform
Facebook	5	5	5	5	5	2	5
Twitter	5	5	5	5	5	2	5
LinkedIn	5	5	5	5	5	2	5
WhatsApp	5	4.5	4	5	1	2	2
Email	5	2	2	5	1	2	2
Websites	5	2	2	5	5	5	2
Slack	5	4.5	5	5	5	2	5
Google forms	5	2	2	4	5	2	1
Skype	5	5	3.5	5	1	2	5
Zoom	5	5	3.5	5	1	2	5

[a] All tools allow easy switching

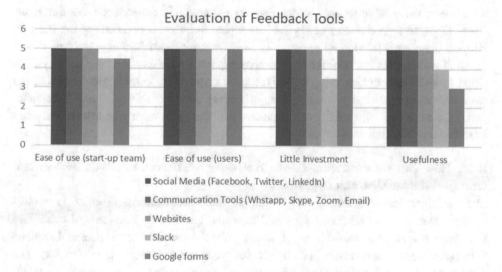

Fig. 5.1 Evaluation of tools against selection criteria (© 2021 IEEE, republished with permission from the publisher)

Further, using these tools, it is possible to take advantage of the customer information which is held by these social networking tools. For instance, the demographic information of customers, their searching patterns across social networks etc., could be useful for startups to identify right customer segment for their product. This however is harder with tools like Zoom or E-mails where you need customer information to be collected explicitly before they could use the tool functionality for interacting with the potential customers.

Websites have higher rating across the utility factor "integration across platforms". This is because the startups can customise easily the website code to integrate them with other feedback tools.

- The feedback acquisition tools are compared across the selection criterias as well (Fig. 5.1). Social networking sites, communication tools like WhatsApp, skype, Zoom and E-mails, and websites score highest across all four criterias i.e., ease of use for both startups and customers, usefulness, and involved investment. Google forms and slack is on average easy to use but need initial effort to set up the interaction model. Slack has lower rating across all factors compared with other technologies.

The startups adopt the feedback technology that is recommended by the experts close to their social networks. Also, direct face to face interaction with customers is considered a routine practice to better explore the markets rather solely depending on online tools. These tools have good chances to be easily adopted in business practices of the startups. Further, these tools provide an opportunity to interact with masses and make data driven decision making. The possibility of incorporating worldwide customers in value proposition innovation activities using online feedback acquisition technologies is one of the research's most noteworthy findings. This is because one of the foundations of customer participation, along with experts and academia, in divergent creativity, which addresses RQ6 and was disseminated in Article 9.

RQ5. How can entrepreneurs assess freelancing platforms to create and sustain strategic relationships with freelancers?

The freelancer has a promising role to play in value proposition innovation as evident by three studies 3, 4 and 5 that answered Research question 3. The startups could find freelancer involvement valuable in exploring potential market for its potential product leveraging across the diversity in competencies processed by freelancer community. The research divides freelancer supported RE activity into three phases—pre-outsourcing, outsourcing and post-outsourcing. Each phase is composed of numerous activities, with each activity either conducted manually by the startup team or using the service provided by the freelancing platforms.

The efforts carried out before to outsourcing the value proposition innovation work to freelancers are referred to as the pre-outsourcing phase. Outsourcing is the stage in which freelancers are chosen to complete work related to value proposition innovation.

The activities beyond outsourcing duties to the chosen freelancer are included in the post-outsourcing phase. This is the stage in which freelancers and startups must communicate to coordinate and provide the anticipated deliverables.

As the number of freelancing platforms are increasing so the complexity in decision-making pertaining to selection of these platforms by the startups. Driven by the classification structure of value proposition innovation, the evaluation framework for selecting freelancing platform is proposed. The framework is composed of 13 factors and not all could be meaningful to all startups. These factors include—Task Selection, Dynamism, Platform selection, Freelancer Evaluation, Freelancer selection and Negotiation, Legal support, Communication, Communication channel selection, Work evaluation, Perspective integration, Dynamic Payment estimation, Usefulness and Ease of use.

The significance of these factors for startup depends on its business state and internal competencies. The European 27 early stage startups in our professional proximity were asked to rate the freelancing platforms like Fiverr, Freelancer, Guru, PeoplePerHour and Toptal and Upwork, as per the evaluation platform. The results indicate the following:

- All platforms are easy to use and perceived to be usefulness to startups.
- Dynamic payments are not supported. In other words, the updation of agreed payment is not possible.
- Perspectives of freelancers and startups cannot be merged online on the freelancing platforms. This maybe because, such platforms are not designed for specially for ideas elicitations and discussions.
- Automatic evaluation of freelancer work/deliverables are not supported. The progress can be checked but evaluation is to be done manually.
- The advice about which communication channels will be best to be used with free-lancers, is not provided by the platforms. The startups can use any channel as per freelancer convenience or startup own preference.
- Platforms provide possibility to have communication with freelancers using different modes like video calls, voice calls, chats, e-mails etc.
- Freelancer recommendation services are provided. The recommendation could provide at least an indication about possible freelancers that could be considered for further selection rounds.
- Legal advice about contract drafting or Intellectual property right issues etc., are not provided by the platforms.
- Multi-criteria based freelancer recommendation is provided to startups. These criterias and great diversity of feedbacks makes recommendation more reliable.
- Platform information is widely distributed to make its unique value proposition publicly available, helping startups to get an initial idea about how such platforms will benefit the startups.
- Platforms does not provide option for modifying the existing contractual conditions.

- Platforms does not provide services to help startups to identify which value proposition innovation activity should be outsourced. This remains internal matter of the startups to make outsourcing vs in-house decision.

The proposed framework is evaluated in real settings by studying the real business impacts of evaluating the freelancing platforms followed by successful outsourcing. This included the financial ratio analysis of 12 startups. Financial statement analysis is a helpful technique for analyzing a company's strength. However, judging the impact of freelancers on financial performance without considering other aspects is a bit subjective. This is because it is harder to measure financial contributions solely made by freelancers because of accurate freelancing platforms evaluations,but it is good enough to quantify the business ramifications.

The financial ratios based on freelancer contribution alone as per judgement of startup team, for three financial years i.e., 2018–2019, 2019–2020 and 2020–2021 indicate strong business strength. This is based on improved values of return on Assets (ROA), return on equity (ROE), Current ratio, Gross Profit Margin, Operating profit margin and Total Asset Turnover ratio.

RQ6. How might academic strategic partnerships be investigated as a means of promoting innovation management?
A review of the book titled "Networks, SMEs and the University: The Process of Collaboration and Open Innovation" aims to strengthen the knowledge about academia and startups partnerships and add to the body of knowledge the critical analysis of the same. The collaborative alliances between Universities and Small and Medium Sized Enterprises (SMEs) had been reported to be the powerful success factor for the startups. However, the collaborations need to be organised across four stages i.e., Identifying strategic needs, accessing, and selecting a partner, implementation, and partnership, and reassessing and reshaping partnerships. Universities could be a power innovation partner, but such partnerships should be established with long term vision and systematically.

The review result shows that successful university and SMEs partnerships as documented in the book in the United Kingdom imply a successful lesson in shattering myths about U-SMEs collaborations in general or the likelihood of their success in particular. This book teaches all SMEs that universities can be valuable knowledge partners and parts of their innovation ecosystem, and that effective collaboration is the consequence of partners working together.

The findings are especially relevant to one type of SME in particular: startups looking to minimize their high market failure rates by increasing their product/market fit. The results indicate that startup community will find collaboration with universities valuable as a crucial external source of expertise for innovation management. This evaluation sparked new research into forming and maintaining strategic ties between academia and startups to encourage value proposition innovation.

Pandemic situations impact the ability of the startups to identify the product features that have match with market needs; the activity that requires direct interaction with the customers at the same physical space. Online tools can overcome this limitation, but early stage startups have too limited resources and lack of access to the potential customers, which make their online interactions quite limited. The divergent creativity is required to identify the Requirement Elicitation methods and tools that could help startups to identify product/market fit with limited same physical space interaction with customers. This had been the basis of the research which involves open innovation involving academia, experts, and researchers to explore potential market for the product. The real results based on consulting experience with Spanish startup which successfully identified its market in pandemic time through blended market research is reported.

The blended market research involves secondary research as well as primary research. Secondary research is conducted using LinkedIn, Secondary reports, Blogs, Client websites using online tools like Zoom, LinkedIn chats and E-mails. The secondary research helped startup to limit the primary research efforts that could otherwise be required to explore the market for its product. This secondary research provided the startup with good details about its potential competitors and customers in the market. Primary research then is conducted to further explore the identified customer segments by digging deeper into their problems with existing solutions and their expectations from new product, heling startup to identify the value proposition.

The primary research involves face to face interactions with the narrowed customer segments through collocated meetings and online communication tools. However, it was observed that using simple prototypes like brochures could be useful in triggering motivation of customers to participate in the process of gathering and validating the value proposition ideas. Finally, the startup could customise their product offering to different customer segments as through the involvement of experts, researchers, academia, and customers, they have well explored the potential market. It was also observed that consulting team spent about 70% efforts on secondary research and remaining 30% on primary research. Out of 30% of primary research efforts, collocated meetings were limited to 40%.

The value proposition innovation process involves generation of diverse creative ideas, validating these ideas and converging them into a solution that could be commercialised. The ability to generate and validate ideas that are both novel and valuable require the involvement of experts and academia with diversity in expertise, knowledge, skills, and background. This strongly depends on the application domain of the startup product, for instance, mobility, health etc., as well as technology underlying the solution, for instance, Artificial intelligence, Cloud, blockchain etc. The academia has rich competencies in different areas of Technology, Management, Social sciences, Law etc., which could help startup not only in identifying unique value proposition but also its implementation and commercialisation. Further, the academia involvement can help startup to overcome the hesitation of the potential customers to participate in innovation activities due to strong brand image of participating universities and the professional and social proximities of academia's with potential customers.

Another research study indicates the academia as a potential and valuable open innovation ecosystem element that could help startup to foster innovation. The two-way knowledge transfer between academia and startups could create synergies between them and could help startups to leverage across the strategic partnerships.

There are increased interests among academic institutions to enter strategic partnerships with startups. This is because of increased focus on entrepreneurship and innovation to build nation economy, strategic university goals for entrepreneurship to achieve sustainability, zeal to gain higher world rankings for the university, increasing financial strength to overcome reduced government financial support etc. Startups could also foster the project based learning in academia by making their business cases accessible to the students, opportunities for internships, access to its disruptive technologies. This helps the academia to build their competencies to apply their learning in an industrial context and gain more competencies and skills to solve real world problems. The startups could take advantage of the competencies owned by the academic in terms of their expertise, knowledge, research etc.

Academia can offer startup with the opportunity to solve their problems using crowdsourcing with its faculty, staff, and students. The eased processes to foster such strategic partnerships in both startup and academia organizational setups will be a good motivator factor for co-innovation. Further, this partnership is easier to be fostered for long-term as it does not get impacted by the "Not Invented Here (NIH)" syndrome that otherwise is common with corporations. The co-innovation will be mutually value adding activity leading to improvement of both academia and the startups.

5.2   Unity and Coherence Between Individual Research Study Outcomes

Research objective is achieved by interrelated 6 research studies that resulted in dissemination of research results in 10 research articles. The research outcomes together answer six research questions thereby meeting 4 sub-research objectives (or overall research objective). The unity is developed as all research outcomes focus on value proposition innovation to improve startup success rates.

Research article 1 investigates the state of value proposition innovation in startups. The results analyse the research trends using classifications and counting of broad research area of value proposition innovation (or requirement engineering) in startups. The research outcomes are disseminated in Article 1. Beside the need for uniform research across all value proposition innovation activities, the result indicate that this area is still growing and there is a need for investing more research efforts in the activity. This sets necessary motivation to plan research protocol for research study 2 and execute the plan. Startups have limited resources and involving external elements of open innovation could help it foster value proposition innovation. The increasing trends across the Gig economy motivate more research efforts to investigate if freelancers could foster innovation and if so, then how such involvement could be leveraged to its full potential. Further, the customer-centric innovation help startups to launch the product as per market needs thereby gain tractions in the market. The role of freelancers and customers become integral to the value proposition innovation in startup context. This triggered the two research studies-research study 2 and research study 4.

Research study 2 analyses the research trends across freelancer involvement across entire value proposition innovation in software startups. The research outcomes is disseminated in Article 2. The research area is promising, but additional research is needed, such as experience reports, case studies, validations, and expert opinions. There are several difficulties in working with freelancers that have been recognized by this research. The research outcomes set motivation for further research in this domain to help startups improve their market success driven by innovation under limited resources. Also, the research outcomes provided the theoretical framework that will help to build conceptual framework in research study 3.

Research study 4 investigates that how global customers can be involved in value proposition innovation through the use of online feedback acquisition tools. The research outcomes are disseminated in Article 6. The study was conducted in pandemic and had great potential to transform the traditional ways of innovating, for instance, co-located meeting with customers to modern ways that involves technology. The result indicates that to involve worldwide customers in value proposition innovation, online feedback acquisition technologies can be deployed. The factors that influence technology adoption are identified, and numerous techniques used during the epidemic are compared to these factors. This triggered the need for research study 6 which investigate the value proposition innovation driven by strategic partnerships with academia and customer involvement.

Research study 3 aims to investigate the freelancer driven innovation practices of startups to explore further about this phenomenon. The research outcomes are disseminated in Article 3,4 and 5.

Article 3 investigates the freelancer driven innovation in startups, identifying the association models, challenges, and associated business impacts. Startups and freelancers could collaborate to drive innovation with positive business outcomes. When used across several innovation efforts over a lengthy time period, the commercial impacts are substantially stronger. The difficulties that such organizations face are also identified.

Research article 4 investigated the value proposition innovation practices of startups after a gap of long time period with respect to the previous study as disseminated in article 3. The case study was re-conducted with focus on value proposition innovation involving same startups as mentioned in article 3 and two young businesses situated in Switzerland and India. This allowed to use the research outcomes of research article 3 as the theoretical background to further investigate the evolved strategies for associations, the ways challenges are addressed, and business impacts reported. These startups were able to address the challenges of involving freelancers in innovation management to undertake value proposition innovation.

The overall research outcome indicates that by involving freelancers, startups can drive value proposition innovation. The startups approach to overcoming the issues of freelancer involvement, as well as the resulting business implications, are discussed. To achieve long-term success, they must form strategic partnerships with freelancers. The outcome of research article 3 and 4 provides necessary theoretical framework for another study which was conducted to analyse how freelancers could foster value proposition innovation in global markets. The results indicate Startups can globalize their operations by enlisting the help of geographically dispersed freelancers for value proposition innovation. The difficulties and business implications of such connections are discussed. Overall, research study 3 indicate the promising role of freelancers in value proposition innovation.

The outcome of research study 3, or article 3,4 and 5, triggered the research study 5. This study based on outcome of articles 3,4 and 5 investigates that how the entrepreneurs could evaluate the freelancing platforms to create and sustain strategic relationships with freelancers? The research outcomes are disseminated in Article 6. The results indicate that Freelancer-supported RE activity is divided into three phases: pre-outsourcing, outsourcing, and post-outsourcing, with 13 variables for evaluating freelancing platforms identified. Financial ratio analysis is used to examine the evaluation framework in real-world situations.

Research study 6 as driven by research study 4, investigates that how strategic partnership with academia and customer involvement could foster value proposition innovation. This study outcomes are disseminated in research article 8, 9 and 10. Research article 8 conducts a review of the book pertaining to startup collaboration with academia in United Kingdom. The outcome indicates that the collaborations between academia and startups in the United Kingdom are a good illustration of such successful collaborations. Such alliances could be formed and maintained in a methodical manner. This provided necessary motivation and framework for further research that resulted in articles 9 and 10.

Article 9 investigate the role of academia and customers in globalisation of startup driven by innovative value proposition. The outcome indicated that blended market research led to the globalization of a Spanish startup in the pandemic. Secondary market research, as well as the backing of specialists, academia, and customers, may not only lead to market success, but may also reduce primary research efforts. Article 10 report that strategic collaboration between academia and entrepreneurs may be a mutually beneficial process. This relationship has the potential to help startups innovate.

All research articles together meet the formulated research objective of the book. They together through missed mode and multiple research methods added to a body of knowledge that how startups could increase their success rates by focusing on value proposition innovation, which is propelled by the involvement of potential consumers as well as other resources such as freelancers and strategic relationships with academia. The research articles have coherence as there is a flow of logic between the articles. They are all connected and provide motivation, theoretical background and research inputs to the articles that follow them in the research process. Figure 5.2 depicts the unity and coherence between individual research outcomes.

5.3   Discussion

At the beginning of this book, the motivation for conducting research in value proposition innovation in software startups context was highlighted. Further, the focus on fostering these innovations with involvement of freelancers, customers, and academia was positioned with respect to theoretical background.

The rational was that startups have higher failure rates (Mullins et al., 2009; Nobel, 2011; Cantamessa et al., 2018; Danarahmanto et al., 2020; Haddad et al., 2020; @@Dvalidze & Markopoulos, 2020; Eesley & Lee, 2020; Rafiq et al., 2021; Santisteban et al., 2021) and mismatch between offered product and market needs is one of the reasons for their failures i.e., faulty value proposition (Alves et al., 2006; Klotins et al., 2015; Giardino et al., 2016; Unterkalmsteiner et al., 2016; Chanin et al., 2017; Cantamessa et al., 2018; Danarahmanto et al., 2020; Carmen, 2021; Rafiq et al., 2021).

The startup community have limited support from body of knowledge as business model innovation literature in startup context is still an underdeveloped issue (Ibarra et al., 2020; Pucihar et al., 2019) as well as value proposition (Antonopoulou & Begkos, 2020; Schmidt & Scaringella, 2020). Further, business model innovations lead to higher performance levels and competitive advantage (Vargo & Seville, 2011; Eggers & Kraus, 2011; Kraus et al., 2020; Albats et al., 2021; Clauss et al., 2021; Latifi et al., 2021). Business model innovation results in value proposition innovation and this value proposition innovation is beneficial for the firm (Antonopoulou & Begkos, 2020; Chandler et al., 2014; Chesbrough, 2007, 2010; Clauss, 2017; Covin et al., 2015; Guo et al., 2021; Rintamäki & Saarijärvi, 2021).

This signifies that investing more research efforts in the value proposition innovation domain could help startups to be more competitive in markets to increase their success rates. Taking advantage of the growing importance of freelancers, innovations in freelancing platforms, technologies to connect with global customers and increasing entrepreneurial inclination of academia, the value proposition innovation could leverage across such developments to provide real value to the startup community. The empirical research conducted through multiple research methods will help startup community to identify innovative value proposition for domestic and global markets thereby overcoming their liabilities of newness and smalless that otherwise hinders their innovation potential. This together with their organizational flexibilities will create synergistic effects in fostering innovations in highly competitive and risky marketplaces. This synergism comes

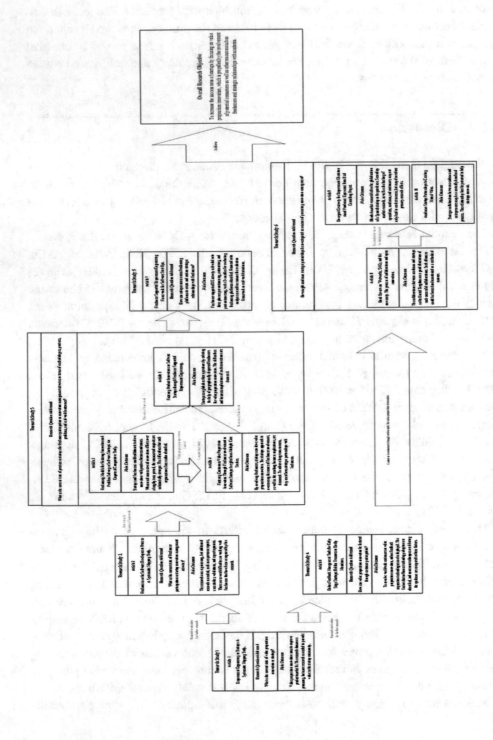

Fig. 5.2 Unity and coherence between individual research outcomes

from their ability to overcome their resource limitations by freelancer, customers, and academia long term trustworthy partnerships and taking best advantage of their flexibilities in adapting to business environment changes.

Business model innovation require series of experimentations in markets to test assumptions about business model to identify scalable and repeatable one (Osterwalder & Pigneur, 2010). The limited resources owned by startups could limit their abilities to conduct these experimentations on larger scale and hence their innovation efforts (Andersen et al., 2022; Arbussa et al., 2017; Berends et al., 2014; Spender et al., 2017).

The research conducted and disseminated in this book suggest that freelancers can be involved across multiple activities of value proposition innovation. This has several positive implications. First, startups have wider options to associate with the freelancers like panel based, task based and hybrid approaches which could be configured as crowd-sourced or non-crowdsourced. The options could be adapted by startups as per the analysis of their internal competencies and external environment analysis. For instance, a startup with too limited financial resources but with technology having great applicability in market, for instance, sanitization technology, could seize the opportunity by outsourcing some of its activities to freelancers. One important activity could be to identify the promising customer segment, for instance, hotels, restaurants etc., which require startups to build prototypes to validate product match with market needs. This task does not require free-lancer to dig deeper into technological aspects and just understanding about functional and non-functional aspects. So, startup could employ crowdsourced task based approach.

In reality, the startups have to interact a lot with customers to build value propositions, which are termed as experimentations. Liability of newness and smallness hinders their ability to reach to global customers through face to face co-located meetings. Further, branding issues could limit their access to global customers perspectives as they could resist in participating in such experimentations even through conducted online. In such situations, freelancers could be successfully integrated with human resources of startups to enlarge their resource base, build internal competencies, for instance, knowledge transfer from freelancers to internal startup team, accessing the freelancer tactical knowledge about customer needs in close proximity to them and implement the acquired value propositions into useful products.

This require long term associations with freelancers so that the knowledge they acquire in previous tasks could be employed to solve challenging problems in future tasks. This long term association is possible, for instance, by creating the panel of such freelancers and providing them with extra benefits, for instance, equity based payments or percentage shared schemes. By configuring outsourcing in a way that the task is outsourced to the team with mix of freelancers and internal startup teams, will help to foster knowledge transfer, helping startup to build their internal competencies. Freelancer's insights or per-spectives about the value could provide another important aspect which otherwise could be missing in the perception build by startup team about the product value proposition.

The monetary benefits help startups to enlarge their panels which help it to reduce its outsourcing transaction cost and innovation costs.

Another important aspect will be their ability to strategically employ the freelancing platform technologies. There are numerous platforms which are continuously innovating to provide values to startups and freelancers. In initial startup operations, creating and maintaining the panel may not be very optimal decision. The access to such platforms could help startup to find global talent (de Peuter, 2011), integrate it with startup team on long term basis and share the benefits for turning him into strategic resource. This is in line with the outcome of the research that reported that freelancers could help businesses to find new capabilities, increasing labour force flexibility, increasing speed to market, and innovating new business models (Adamson, 2021). The global diversity of freelancers across these platforms will help startup to get foreign market information as well as access to global customers based on globally distributed freelancers.

The globalisation is fostered through innovative value proposition through freelancer involvement as evident through the conducted research of research study 3. As the number of studies pertaining to startups and value proposition innovation are too limited, the case studies conducted in the research study 3, provide a good indication that freelancers could help startup foster value proposition innovation with meaningful business outcomes.

The customers are the main stakeholders that influence the innovation activities. Co-creation with customers is the main factor responsible for having an innovative value proposition. Globalisation results in opportunities for market expansion, reduction in production costs, gaining access to scientific and technical resources and much more. Startups have limited resources to have access to global customers and hence is one of the factors that limit their abilities to globalise. Software startups can distribute their products through channels like APP stores or through websites.

The major challenge to have a scalable and repeatable business model is to formulate an innovative value proposition, for which customer participation is mandatory. The freelancer's involvement could help startup to get their perspectives from customer point of view based on their understanding of customers in close proximity to them. However, deeper insights are only brough through interactions with customers. In pandemic, co-located interactions are restricted and lack of branding as well as innovativeness of the product limit the customer motivation to participate in such interactions. This could however be handled in various ways.

First, through the participation of academia, experts and freelancers, the startup could narrow down to meaningful customer segments. This will help them to focus their efforts on only early adopters. The reputation and proximities of experts and academia could help startup to make their professional proximities closer with customers.

Second, leveraging across the innovations across the feedback acquisition technologies like social networking sites, could help startup to raise awareness, interests, desires, and actions on part of customers about the product. These technologies help startup to make their professional proximities with customers closer. This help startup to access global markets and globalise their business operations. The increased federal government focus on entrepreneurship, university focus on entrepreneurship and third mission for social and economic contribution (Compagnucci & Spigarelli, 2020) will make academia more aligned with strategic partnership with startups.

In continuously changing software industry, the software startups work under higher uncertainties. Adopting to these uncertain fluctuations require startups to undertake business model innovations which strongly depend on their dynamic capabilities (Teece, 2018; Čirjevskis, 2019) as well as value proposition innovations. Dynamic capabilities require startups to renew their existing competencies (Schriber & Löwstedt, 2020). The startups have limited strategic resources which means that the main source of their dynamic capabilities is their flexibility in organisational setup (Fabrizio et al., 2021). Dynamic capabilities help startups to have sustained competitive advantage. This theory is used to explain sustained competitive advantage in the fluctuating business environments. So, it could be used as the basis to predict if startups could gain dynamic capabilities from strategic involvement of freelancers, customers, and academia in value proposition innovation activities.

The firms survive in the market only when their product is perceived valuable by the customers. The product of service is innovative or not, is decided by customers only. Synchronising product value proposition with the market need is the reason for startup success. Ability to sense customer needs, mobilise resources to turn identified opportunities into new products and transforming the resource base strongly depend on ability to co-create with customers. Online feedback acquisition tools like social media helps startups to reach to masses globally to make customers aware of their product, get validation about value proposition hypobook and finally deliver the innovation to market. Such tools allow two-way interaction between startups and customers. This interaction is further enhanced with freelancer and academia participation in value proposition innovation activities.

Streamlining the customer involvement for co-creation definitely positive impacts the dynamic capabilities of the startups. This allows them to easily identify opportunities, seize opportunities and transform resource base and competencies. Technology adoption, support of customers, freelancers and academia makes learning and exploiting new knowledge much easier which strongly impacts the dynamic capabilities. This is in line with one of the observations reported in (Teixeira et al., 2020) that startup growth is positively impacted by their dynamic capability to sense user needs. The customer involvement is central for their ability to learn from customers about their needs. Strategic co-creation with customers will further strengthen dynamic capabilities.

Involvement of freelancers will help startup to strengthen its sensing, seizing, and transforming dynamic capabilities. This is because freelancers help startups to learn more about global customer needs. Further, in startups investigated in the research study indicated that joint working of startup team and freelancers help startup team to learn from freelancers and vice versa. This result in knowledge sharing that help startups build their competencies that they otherwise lack in-house.

Identifying opportunities require startups to mobilise their resources but limited resources sometimes make this practically impossible. Here, by outsourcing fewer valuable activities to freelancers, for instance through task-based approach, the startup team could be allocated to more valuable tasks.

Further, if startup employs a panel based approach, the long term association with freelancers will help startup to decide which freelancers should be mobilised and integrated with the startup team for new challenging task. As the size of the internal team and freelancer panel grows, the transformation becomes easier. Thus, freelancer involvement helps startup build dynamic capabilities.

Involvement of academia will positively impact startup dynamic capabilities. Long term associations with academia will help startup get an access to expertise residing in academia for building their dynamic capabilities of sensing, seizing, and transforming.

Value proposition innovation is a tool for competitive advantage, but the long term competitive advantage depends on its continuous innovations. The ability to innovate value proposition continuous require startups to overcome their liability of newness and smallness for which open innovation with freelancers, customers, and academia in a strategic way is required. This strengthens the dynamic capabilities of the startups in making a response to changing business environment through its value proposition innovation and gain sustained competitive advantage.

The involvement of freelancers, customers, and academia helps startups overcome their resource limitations. This together with their organizational flexibility will further strengthen their dynamic capabilities. This synergism however depends on the ability of startups to create and maintain strategic associations with them. The startup needs to formulate within business strategies, the strategies to create and maintain associations with freelancers, customers, and academia.

5.4 Book Contribution

This section highlights the major contributions made by this book to theory, methodology and the practice.

(a) Theoretical Contribution

The contribution of the book to the body of knowledge is twofold. First, advancing the body of knowledge and second, providing new ways of fostering value proposition innovations in a highly dynamic environment than what was being done traditionally. The body of knowledge is enhanced in several ways.

First, advancing the literature on value proposition innovation in software startup context. Second, advancing the literature on open innovation to foster value proposition innovation driven by freelancers, customers, and academia. Third, advancing the body of knowledge that pertains to business model innovation activities in times of pandemic. Fourth, using the dynamic capability lens, adding to body of knowledge the value proposition through freelancers, customers, and academia involvement that strengthens startup dynamic capabilities and leads to sustained competitive advantage. In particular, the results advance the understanding of how startups could find, implement, and evolve product value propositions in highly dynamic environment by strategic associations with freelancers, customers, and academia. This will provide startup community empirical evidence that how they could overcome their liabilities of newness and smallness to strengthen their dynamic capabilities leveraging across their flexibilities in conducting business operations.

The new ways of fostering value proposition innovation as compared to the way it was being done traditionally. First, possibility of outsourcing the core business operations, for instance, value proposition activities to freelancers. However, the strategic association and integration of freelancers with startup team is required to get maximum business benefits. Second, possibility of customer involvement in co-creation process through online feedback acquisition tools. Traditionally, this had been done through face to face co-located meetings with customers replying mostly on interviews, observations etc. Traditional ways limit the spatial access of the startup team, with focus on selected geographical areas, mostly domestic niche markets. Third, possibility of integrating academia, freelancers, other experts, and customers in co-creation activities. This helps startup to get maximum market insights with minimum efforts.

The book builds up the literature of value proposition innovation pertaining to both management and technical sciences. Although, the findings are more management oriented but technical researchers will find them useful to bridge the gaps through their technical innovations, for instance, computationally advanced algorithms or tools. The outcome thus will foster two-way knowledge transfer between managers or entrepreneurs and technical researchers.

(b) Methodological Contribution

Another contribution of this book is methodological innovation. As the research domain of value proposition innovation is limited so making theoretical and practical contributions is possible through the use of multiple research methods involving systematic mapping studies, case studies, surveys, reviews, and experience reports. The methodological contributions have several dimensions.

First, through multiple mapping studies the outcomes were conceptualised to scope the case studies that follow. Second, multiple case studies were conducted which helped to find generalisability across the studied phenomenon. The results were verified with member checking thereby ensure their reliability. Third, experience reports were used to include practical experiences to explore the research domain better. The use of multiple research methods helped to include different perspectives about the research domain and make results more valid and reliable. Fourth, the research study involved real research settings. For instance, the case studies involved study of phenomenon in the real context of startups. Using interviews, site visits and observations helped to provide more realism to the research. Further, experience reports reflect the real practical experiences of the author in different projects. Fifth, the cases involved in case study are the ones that have gained good market share and are growing fast besides also including some younger startups.

As limited startups employ freelancers for value proposition on larger scale, the studied startups are good representative of population sample. Further experience reports represent the practical experiences across multiple projects and hence more generalisable knowledge. Sixth, use of different types of data-qualitative and quantitative makes the study results more informative.

(c) **Practical Contribution**

The book outcomes have several practical contributions as well.

First, entrepreneurs could use the case study outcomes and adopt them as per their own working context, internal competencies, and external opportunities. The empirical evidence could be readily adopted in startup working practices. Second, the customer involvement in co-creation using feedback tools provides a good solution to entrepreneurs. All startups want to globalise their operations. The online feedback acquisition technology can be easily adopted by startups for connecting with globally distributed customers. Third, the freelancing platform evaluation framework will help startups to take advantage of growing number of freelancing activities across these platforms into more strategic associations leading to more dynamic capabilities. Fourth, freelancers could associate with those startups that provides them value with focus for long term without compromising their working autonomy. Fifth, academia could also find growing associations of startups with freelancers, customers, and their peer academic institutions as the metric for their valuation and future prospects. This will help them build more confidence in establishing long term associations with them. Sixth, for customers, the outcome is of great importance. The ongoing collaborative operation of startup gives an indication of the promising innovations that will be commercialised. This will help them in real to decide if they should participate in co-creation or not.

The outcomes will help startups to foster value proposition innovations and business model innovations in highly uncertain markets by continuously strengthening their dynamic capabilities.

The outcomes are likely to bring great value to startups if they do the following. First, make all associations with customers, freelancers, and academia strategic. To accomplish this, they must focus on mutual values, helping their partners perceive value from such associations. Second, they should always maintain their business flexibility and continuously integrate customers, freelancers, and academia in their business operations for knowledge generation and exploitation. Knowledge exploitation signify the building of competencies by putting the acquired knowledge in practice within the organisation. Further, working in joint teams makes knowledge sharing and learning a source of dynamic capabilities.

Thus, entrepreneurs should focus on building their dynamic competencies not only based on their flexibilities but also on their enlarged resource based, resource base with strategic involvement of freelancers, customers, academia, and startup team.

References

Adamson, C. (2021). How freelance developers can help your midsize business. *Harvard Business Review*.

Albats, E., Podmetina, D., & Vanhaverbeke, W. (2021). Open innovation in SMEs: A process view towards business model innovation. *Journal of Small Business Management*, 1–42. https://doi.org/10.1080/00472778.2021.1913595

Alves, C., Pereira, S., & Castro, J. (2006). A study in market-driven requirements engineering. In *Proceedings of the 9th Workshop on Requirements Engineering (WER '06)*, Rio de Janeiro, Brazil.

Andersen, T. C. K., Aagaard, A., & Magnusson, M. (2022). Exploring business model innovation in SMEs in a digital context: Organizing search behaviours, experimentation and decision-making. *Creativity and Innovation Management, 31*(1), 19–34. https://doi.org/10.1111/caim.12474

Antonopoulou, K., & Begkos, C. (2020). Strategizing for digital innovations: Value propositions for transcending market boundaries. *Technological Forecasting and Social Change, 156*, 120042.

Arbussa, A., Bikfalvi, A., & Marquès, P. (2017). Strategic agility-driven business model renewal: The case of an SME. *Management Decision, 55*, 271–293. https://doi.org/10.1108/MD-05-2016-0355

Berends, H., Jelinek, M., Reymen, I., & Stultiëns, R. (2014). Product innovation processes in small firms: Combining entrepreneurial effectuation and managerial causation. *Journal of Product Innovation Management, 31*, 616–635. https://doi.org/10.1111/jpim.12117

Cantamessa, M., Gatteschi, V., Perboli, G., & Rosano, M. (2018). Startups' roads to failure. *Sustainability, 10*(7), 2346. https://doi.org/10.3390/su10072346

Chandler, G. N., Broberg, J. C., & Allison, T. H. (2014). Customer value propositions in declining industries: Differences between industry representative and high-growth firms. *Strategic Entrepreneurship Journal, 8*(3), 234–253.

Chanin, R., Pompermaier, L., Fraga, K., Sales, A., Prikladnicki, R. (2017). Applying customer development for software requirements in a startup development program. In *Proceedings of the 2017 IEEE/ACM 1st International Workshop on Software Engineering for Startups (SoftStart)*, Buenos Aires, Argentina, 21 May 2017, pp. 2–5.

Chesbrough, H. (2007). Business model innovation: It's not just about technology anymore. *Strategy & Leadership, 35*(6), 12–17.

Chesbrough, H. (2010). Business model innovation: Opportunities and barriers. *Long Range Planning, 43*(2), 354–363.

Čirjevskis, A. (2019). The role of dynamic capabilities as drivers of business model innovation in mergers and acquisitions of technology-advanced firms. *Journal of Open Innovation: Technology, Market, and Complexity, 5*(1), 12. https://doi.org/10.3390/joitmc5010012

Clauss, T. (2017). Measuring business model innovation: Conceptualization, scale development, and proof of performance. *R&d Management, 47*(3), 385–403.

Clauss, T., Breier, M., Kraus, S., Durst, S., & Mahto, R. V. (2021). Temporary business model innovation–SMEs' innovation response to the Covid-19 crisis. *R&D Management.* https://doi.org/10.1111/radm.12498

Compagnucci, L., & Spigarelli, F. (2020). The Third Mission of the university: A systematic literature review on potentials and constraints. *Technological Forecasting and Social Change, 161.* https://doi.org/10.1016/j.techfore.2020.120284

Covin, J. G., Garrett, R. P., Jr., Kuratko, D. F., & Shepherd, D. A. (2015). Value proposition evolution and the performance of internal corporate ventures. *Journal of Business Venturing, 30*(5), 749–774.

Danarahmanto, P. A., Primiana, I., Azis, Y., & Kaltum, U. (2020). The sustainable performance of the digital start-up company based on customer participation, innovation, and business model. *Business: Theory and Practice, 21*(1), 115–124.

Dvalidze, N., & Markopoulos, E. (2019). Understanding the nature of entrepreneurial leadership in the startups across the stages of the startup lifecycle. In *International Conference on Applied Human Factors and Ergonomics* (pp. 281–292). Springer.

de Peuter, G. (2011). Creative economy and labor precarity: A contested convergence. *Journal of Communication Inquiry, 35*(4), 417–425. https://doi.org/10.1177/0196859911416362

Eesley, C. E., & Lee, Y. S. (2020). Do university entrepreneurship programs promote entrepreneurship? *Strategic Management Journal, 42*(4), 833–861. https://doi.org/10.1002/smj.3246

Eggers, F., & Kraus, S. (2011). Growing young SMEs in hard economic times: The impact of entrepreneurial and customer orientations—A qualitative study from Silicon Valley. *Journal of Small Business & Entrepreneurship, 24*, 99–111.

Fabrizio, C. M., Kaczam, F., de Moura, G. L., da Silva, L. S. C. V., da Silva, W. V., & da Veiga, C. P. (2021). Competitive advantage and dynamic capability in small and medium-sized enterprises: a systematic literature review and future research directions. *Review of Managerial Science*, 1–32. https://doi.org/10.1007/s11846-021-00459-8.

Giardino, C., Paternoster, N., Unterkalmsteiner, M., Gorschek, T., & Abrahamsson, P. (2016). Software development in startup companies: The greenfield startup model. *IEEE Transactions on Software Engineering, 42*, 585–604.

Guo, H., Yang, J., & Han, J. (2021). The fit between value proposition innovation and technological innovation in the digital environment: Implications for the performance of startups. *IEEE Transactions on Engineering Management, 68*(3), 797–809. https://doi.org/10.1109/TEM.2019.2918931

Haddad, H., Weking, J., Hermes, S., Böhm, M., & Krcmar, H. (2020). Business model choice matters: How business models impact different performance measures of startups. In *Wirtschaftsinformatik (Zentrale Tracks)* (pp. 828–843).

Ibarra, D., Bigdeli, A. Z., Igartua, J. I., & Ganzarain, J. (2020). Business model innovation in established SMEs: A configurational approach. *Journal of Open Innovation: Technology, Market, and Complexity, 6*(3), 76. https://doi.org/10.3390/joitmc6030076

Klotins, E., Unterkalmsteiner, M., Gorschek, T. (2015). Software engineering knowledge areas in startup companies: A mapping study. In *Proceedings of the International Conference of Software Business*, Braga, Portugal, pp. 245–257.

Kraus, S., Clauss, T., Breier, M., Gast, J., Zardini, A., & Tiberius, V. (2020). The economics of COVID-19: Initial empirical evidence on how family firms in five European countries cope with the corona crisis. *International Journal of Entrepreneurial Behavior & Research, 26*, 1067–1092.

Latifi, M. A., Nikou, S., & Bouwman, H. (2021). Business model innovation and firm performance: Exploring causal mechanisms in SMEs. *Technovation, 107*. https://doi.org/10.1016/j.technovat ion.2021.102274

Mullins, J., Mullins, J. W., Mullins, J. W., & Komisar, R. (2009). Getting to plan B: Breaking through to a better business model. Harvard Business Press.

Nobel, C. (2011). *Teaching a 'lean startup' strategy*. HBS Working Knowledge, pp. 1–2.

Osterwalder, A., & Pigneur, Y. (2010). *Business model generation: a handbook for visionaries, game changers, and challengers* (Vol. 1). Wiley. ISBN: 978-0-470-87641-1.

Pucihar, A., Lenart, G., Borštnar, M. K., Vidmar, D., & Marolt, M. (2019). Drivers and outcomes of business model innovation-micro, small and medium-sized enterprises perspective. *Sustainability, 11*, 344. https://doi.org/10.3390/su11020344

Rafiq, U., Melegati, J., Khanna, D., Guerra, E., & Wang, X. (2021). Analytics mistakes that derail software startups. In *Evaluation and Assessment in Software Engineering* (pp. 60–69). https://doi. org/10.1145/3463274.3463305

Rintamäki, T., & Saarijärvi, H. (2021). An integrative framework for managing customer value propositions. *Journal of Business Research, 134*, 754–764.

Santisteban, J., Mauricio, D., & Cachay, O. (2021). Critical success factors for technology-based startups. *International Journal of Entrepreneurship and Small Business, 42*(4), 397–421. https:// doi.org/10.1504/IJESB.2021.114266

Schmidt, A. L., & Scaringella, L. (2020). Uncovering disruptors' business model innovation activities: Evidencing the relationships between dynamic capabilities and value proposition innovation. *Journal of Engineering and Technology Management, 57*, 101589. https://doi.org/10.1016/j.jen gtecman.2020.101589

Schriber, S., & Löwstedt, J. (2020). Reconsidering ordinary and dynamic capabilities in strategic change. *European Management Journal, 38*(3), 377–387. https://doi.org/10.1016/j.emj.2019. 12.006

Spender, J.-C., Corvello, V., Grimaldi, M., & Rippa, P. (2017). Startups and open innovation: A review of the literature. *European Journal of Innovation Management, 20*(1), 4–30. https://doi. org/10.1108/EJIM-12-2015-0131

Teece, D. J. (2018). Business models and dynamic capabilities. *Long Range Planning, 51*(1), 40–49. https://doi.org/10.1016/j.lrp.2017.06.007

Teixeira, E. G., Moura, G. L. D., Lopes, L. F. D., Marconatto, D. A. B., & Fischmann, A. A. (2021). The influence of dynamic capabilities on startup growth. *RAUSP Management Journal, 56*, 88–108. https://doi.org/10.1108/RAUSP-08-2019-0176

Unterkalmsteiner, M., Abrahamsson, P., Wang, X., Nguyen-Duc, A., Shah, S. Q., Bajwa, S. S., ... & Yague, A. (2016). Software startups—A research agenda. *e-Informatica Software Engineering Journal, 10*(1), 89–123.

Vargo, J., & Seville, E. (2011). Crisis strategic planning for SMEs: Finding the silver lining. *International Journal of Production Research, 49*, 5619–5635.

Conclusion and Strategic Directions

Summary points about academic, freelancing, and customer-driven value proposition innovations that are more long-lasting are provided in this chapter to wrap up the subject. Dynamic capabilities play a clear role in having ongoing value proposition developments. Various abilities come into play when these components of the open innovation ecosystem are strategically incorporated into innovation activities. Startups can achieve lasting competitive advantage and market innovation through the use of these competencies. The startup community will finally find useful the strategic directions that are focused on dynamic capabilities to achieve sustainable competitive advantage.

6.1 Concluding Remarks

Based on their organizational flexibility and capacity to leverage their collaborations with academia, freelancers, and customers, startups can establish dynamic skills to continuously improve their value propositions in changeable business conditions. The majority of the discoveries in this book are based on actual startup practices in pandemic, so they can teach their peers valuable lessons outside of pandemic. This book focuses on analyzing how entrepreneurs innovated their value propositions in the pandemic and how these innovative methods may contribute to startups' dynamic capacity to constantly evolve. The pandemic has presented difficulties for startups doing its typical business operations. Successive pandemic experiments to adapt business strategies to pandemic scenarios will serve as the foundation for lessons learned for the new normal. The following are some insightful conclusions for readers of the book.

© The Author(s), under exclusive license to Springer Nature Switzerland AG 2022
V. Gupta, *Strategic Value Proposition Innovation Management in Software Startups for Sustained Competitive Advantage*, Synthesis Lectures on Technology, Management, & Entrepreneurship, https://doi.org/10.1007/978-3-031-18322-5_6

(a) By involving freelancers, startups will be better able to recognize, seize, and transform business opportunities. This is so that entrepreneurs may better understand the demands of customers around the world thanks to freelancers. Additionally, the research study's investigation of businesses revealed that collaboration between startup teams and freelancers helps startup teams learn from contractors and vice versa. Sharing of knowledge allows entrepreneurs to develop the competencies they would otherwise lack internally. The startup will be able to choose which freelancers should be mobilized and integrated with the startup team for new hard tasks with the support of its long-term engagement with freelancers. The change gets simpler as the internal staff and freelancer panel expand. Involving independent contractors enables startups to develop dynamic skills.

(b) Strategic partnerships with academic institutions will give startups access to the know-how they need to recognize, grab, and transform business opportunities. Academic research on developing market opportunities and how to seize them may offer businesses helpful guidance. A strong tool for startups to get knowledge about the global market may be the ongoing relationships between academic institutions. Additionally, academia might support businesses in repurposing and mobilizing their resources to market their novel ideas on continuous basis.

(c) In order to understand the needs of their customers, they must involve the customers. Dynamic skills will be strengthened even more through strategic co-creation with clients. Customers could advise the startup of their upcoming demands and those that are still unfulfilled. All innovation endeavours must include customers. Their participation aids startups in precisely determining their future course and allocating their resources accordingly.

   Finally, their ongoing assistance will enable entrepreneurs to continuously innovate, giving them a sustainable competitive advantage. Streamlining consumer participation for co-creation has a good impact on startups' dynamic skills. They are able to easily see opportunities, seize opportunities, and transform their skill set and resource base as a result.

Adoption of technology, customer, freelancer, and academic assistance make learning and using new information much easier, which has a significant impact on dynamic skills. But for this to happen, startups must publicly and strategically innovate using these components of the innovation ecosystem. The dynamic capability beyond pandemic will benefit from greater technical advances, more customer and business community use of technologies, and increased academic concentration on their "third mission." The end outcome will be more market innovations, a stronger economy, and higher startup success rates.

6.2   Strategic Directions

This section outlines various strategic directions to assist startups in enhancing their capacity to openly innovate continually in varying markets. These strategic directions are divided into three groups: relationships with customers, freelancers, and academic institutions.

(a) Associations with Freelancers

Startups must incorporate ongoing experimentation into the definition of their business models as well as their collaboration with independent contractors. The partnership proposal should be founded on shared principles that enable startups to address their immediate and long-term business issues as well as support independent contractors in achieving their career goals. The biggest difficulty arises during the early phases of a startup's life cycle, when resources, including expertise, are insufficient to determine future business needs, choose freelancers, and decide whether to outsource certain tasks. Startups may not have a fantastic financial offer for independent contractors at this time, but partnerships may be cultivated based on participation in business revenues and potential career progression. However, by carefully choosing freelancing sites, one can engage qualified freelancers with little effort, eventually leading to long-term collaborations. The entrepreneurs must also understand that the long term should take precedence over short-term business indicators when determining the success of such relationships. In other words, companies must establish a variety of business success measures, including short- and long-term, monetary and non-monetary ones.

Using encounters with freelancers as opportunities for mutual learning is another factor to take into account when creating strategic alliances. Internal team competences will be developed as a result, and freelancers will be included in a variety of company operations, including core functions. By doing this, the market insights are presented by various information sources, such as freelancers and startup teams, and rigorously validated from many angles. This aids in the rationalization of market perception and the alignment of business models for startups.

(b) Associations with customers

Social media channels should be used to encourage interactions with clients because they are where people are most active these days. Additionally, social media creates strong relationships between users of social networks on both a personal and professional level, which will aid entrepreneurs in gaining logical market insights and improving communications about their products.

In order to communicate with customers, it is necessary to strategically adopt technologies, as both these technologies and the patterns of client adoption are changing rapidly. It should be simpler for startups to adopt new technologies in accordance with the preferences of their customers since the user experience with one technology can typically be translated to another with minimum self-training.

Additionally, because they have limited resources, startups should prioritize finding the high-paying clients before attempting to serve the requirements of a broad and sizable group of people. In other words, startups should refrain from identifying more value propositions from various client niches. Instead, the goal should be to first satisfy a certain consumer segment before gradually increasing market share.

(c) Associations with academia

Startups could obtain a wealth of resources from academia, such as market insights, assistance with commercialization, freelancers, etc. However, they should strategically integrate with the academic pillars of teaching and research in order to have more profound and strategic ties with them. Startups will be assisted in finding solutions to their business difficulties, for instance, through ongoing contacts with students through case teaching approaches. Additionally, they will be able to recognize academic talent that has the potential to be integrated as a freelancer and then an employee. Startups could experience a paradigm shift as a result of the rigorous academic research available to them, for example by incorporating it into new business models. To successfully develop their business operations, startups must adopt the continuous experimentation procedures to gather continuous learnings.

Furthermore, it is important to comprehend how a company could use the branding resources of academic institutions to leverage its branding-based value proposition innovation difficulties. Collaborations between universities may enable entrepreneurs to easily obtain information about the global market and gain access to clients there.

By continuously enhancing their dynamic capabilities, the strategic directions will assist startups in fostering value proposition and business model breakthroughs in extremely uncertain markets. If they carry out the following, these will probably be of tremendous help to startups. Make all relationships with clients, freelancers, and academia strategic first. They must emphasize shared ideals in order to achieve this, assisting their partners in seeing the value in these relationships. Second, they should regularly include clients, independent contractors, and academia into their business operations for knowledge creation and exploitation. Knowledge exploitation refers to developing competencies inside an organization by applying newly learned knowledge. Working in collaborative teams also makes learning and information exchange a source of dynamic skills.

Entrepreneurs should therefore concentrate on developing their dynamic competencies by strategically involving freelancers, clients, academia, and startup team in addition to basing them on their flexibility and expanded resource base.

Appendix: Case Report and Teaching Case

Case Report

Turning Hurdles into Opportunities Strategically: Customer, Academic, and Freelancer Support

Note: *The startup and the people referenced in the case are real, but their names have been changed to protect their anonymity.*

Peter and James, cofounders of the startup "SocietyTech", were reflecting on their seven years of entrepreneurship. SocietyTech is a social startup that focuses on solving social problems using technical advancements. They recalled how they founded their business in a room of their home with only the two co-founders as employees and with insufficient funding. They talked about the difficulties they had when starting a business in the economic-social sector, how they overcame those obstacles, and what valuable lessons they learned along the way. So how can a startup that operates in a industry with too limited resources especially funding, like the social sector of economy, as well as those in other sectors, innovate successfully? How might they overcome the drawbacks of being new and small? How could they successfully openly innovate with their limited resources?

About SocietyTech

Social sector is defined as "*the part of the economy that includes the activities for social good, i.e., providing benefits to the society by addressing the social problems like poverty, health issues, lack of education, hygiene, hunger,* etc." (Gupta, 2021a). There are many issues facing society, and both the government and non-governmental organizations, for

V. Gupta, *Strategic Value Proposition Innovation Management in Software Startups for Sustained Competitive Advantage*, Synthesis Lectures on Technology, Management, & Entrepreneurship, https://doi.org/10.1007/978-3-031-18322-5

instance charitable trusts, start-ups for social enterprises, are working to find the best ways to support social innovation. The term "social innovation" describes the process of enhancing the social sector by introducing novel products and services into the community. Startups are regarded as the market's suppliers of cutting-edge goods. The social sector might benefit from their ideas. The variety of intended consumers of the inventions, however, makes it more difficult for businesses to implement social innovation. For instance, the demographics, expectations, habits, technical experience, and geographic reach of citizens as users are too different (Gupta, 2021b).

An online startup called SocietyTech seeks to use technical advancements to address social issues. According to their corporate strategy, the social issues that need to be resolved are either found by their own team or as a result of requests from authorities, non-governmental organizations, or even other social institutions. In either scenario, the proposals are assessed according to their viability and impact. Projects having bigger benefits are expected to be more appealing to funders, such as government funding and corporate social responsibility funds. The social sector generates revenue from funders rather than from citizens directly since donors have a responsibility to assist citizens who are experiencing social problems. Additionally, the amount of revenue earned does not allow SocietyTech to make significant profits. Through cut expenses, business choices must be made as efficiently as possible from the time a business idea is developed to the point of effective commercialization. SocietyTech will be able to sustain its company operations by increasing product or service offerings and lowering associated costs.

Since its founding in 2016, SocietyTech has had significant financial difficulties, but by putting excellent methods into practice, it has turned all of these challenges into possibilities. In 2022, this business is no longer a startup and is successfully operating in numerous nations.

Challenges During Early Stages of Startup Life Cycle

Unlike other early-stage entrepreneurs, the startups encountered additional difficulties when running their businesses. The difficulties come from both those that are unique to the startup community and those that come from difficult industries, particularly those engaged in social good. This startup has a finite amount of funding as well as human resources. The business owners had no prior startup or social innovation expertise. Because of their technological skills, they could successfully translate commercial concepts into products. Funding was challenging because of the widespread belief that governments alone are responsible for delivering social services. Additionally, it was more difficult to secure funding due to the lack of resources available to support social goods. Businesses did make offers as part of their social corporate responsibility, but entrepreneurs thought that it required certain professional connections, prior success with successful launches, and the capacity to demonstrate the innovative potential of the offered good or service.

Additionally, it is preferable to undertake market research through contacts with social institutions, individuals, and governments in order to successfully introduce creative products. The businesses found it challenging to develop their value propositions and validate them due to the bureaucracy in public organizations and the diversity of citizens.

Startup planned to cut human labor costs by outsourcing work to independent contractors to get around some of these restrictions, although this was originally more challenging. This was due to the fact that startups don't have much to give independent contractors to entice them to collaborate for a longer period of time. For instance, their initial plan to give a "percentage share" of future corporate profits proved to be unsuccessful for two reasons. First, freelancers were unable to recognize the upstart companies' innovative potential. Second, they hold the opinion that startups are doomed to failure and that it is difficult to put your faith in them for future profits.

Another issue was that people were uninterested in participating since they couldn't see how the solution would resolve their current issues. Additionally, they demanded compensation for their engagement, which was more difficult for companies to provide. Additionally, they have little faith in startups because branding was a big factor in fostering participation motives.

Strategies to Suceed

To successfully innovate in the market, the startup had to continually test out its tactics. As was mentioned in the previous section, it first encountered a number of difficulties. Before generating ideas for potential solutions, entrepreneurs with technical knowledge need deeper understandings of the social sector area. The startup's technical skills were a benefit because they allowed them to take on some implementation and post-feedback driven evolution activities. The firm chose to collaborate with academic universities, taking use of their skills and the chances available there.

Due to close professional relationships between the university professors and the entrepreneurs' homes, the partnership was made simple. Additionally, universities are putting more of an emphasis on encouraging entrepreneurship and achieving their "third purpose" The startup's emphasis on social innovation fit perfectly with universities' "third purpose." The agreement that 20% of startup's net profits from sales will be paid to the university helped to strengthen the relationship with colleges. Universities must increase their financial resources to support social innovation initiatives that need finance, even though this was a social innovation and the returns won't be great. This collaboration aided the startup in the following ways:

- Having access to university resources including academic experts, research expertise, student interns, etc.

- Aided startup in acquiring university branding to foster confidence among social innovation stakeholders, including the government, public institutions, clients, and freelancers.
- The ability to address social institutions since universities have easy access to them.
- To address societal issues, university social campaigns were established in several areas of the region. Entrepreneurs that participated in social campaigns had the chance to investigate the problem domain through interviews and observations. This also aided it in developing closer professional ties with neighborhood NGOs, which enabled it to investigate problem domains and approach possible users of its social innovations. With the support of these campaigns, the firm was able to connect with the public and engage them in value proposition innovation activities.

The firm was able to strengthen its local social presence thanks to the university's assistance. Increased social activities and a connection with a university helped it improve its reputation. This increased the level of trust between startups and independent contractors. Both groups of freelancers—one from the university campus and the other from freelance platforms—were chosen. Selections from sites for freelancing were chosen after carefully examining prior reviews and skills listed by freelancers. Because entrepreneurs have strong technical competencies and strong backing from academic specialists, the freelancer evaluation was practical. Three components made up the startup's overall proposal: (a) Payment for completed work; (b) Career-growth opportunities, such as Industrial Doctorates through partner universities; and (c) Business profit sharing in proportion to freelancer participation. The startup compiled a list of potential freelancers, and those chosen were drawn both from this panel and outside sites. Startup stopped searching for freelancers on other platforms after it had a sizeable enough part of the market and a platform with enough qualified freelancers to update it. But, new freelancers are always added to the platform owned by it based on referals, requests and professional relationships with startup team. Nearly all of the functionality listed in (Gupta et al., 2022) are present on the platform that the startup maintains.

Business Impact of Strategies

The existence of SocietyTech in numerous countries and its financial stability demonstrate the company's business success after seven years of effective operation (Table A.1).

Dynamic Capabilities and Current Strategies

Diverse freelancers make up the startup's freelancer panel. Their varied backgrounds account for the diversity. Culture, aptitude, and nation of residency. This variety aided

Table A.1 Financial ratios of SocietyTech

Ratios	Financial ratios (2020–2021)	Overall impact
ROA	23	Higher profits relative to the assets
ROE	29	Higher profitability relative to the stockholder equity
Current ratio	1.65	Improved business liquidity
Gross profit margin	39	A rise in profitability as a result of rising net sales and falling cost of goods sold
Operating profit margin	21	Profitability growth and lower operating expenses (variable costs)
Total asset turnover ratio	1.8	A higher value denotes rising revenues relative to average assets

startups in enlisting international freelancers to investigate international markets. As an example, a freelancer from Greece gave a startup deep insights into some of the societal issues in their neighboring nations and the remedies already in place. Startups were given initial indicators about the available business potential by this information. This is feasible since a startup always has an open call for new idea submissions on its platform. Freelancers are welcome to offer their thoughts on any topic, including future opportunities, innovative ways to improve current company procedures, and other topics. They are given a consistent percentage share of the company's income in the event of successful commercialization. Additionally, freelancers are tasked with turning business concepts into software solutions. Entrepreneurs were able to collaborate with independent contractors and cut the costs associated with miscommunication and resulting errors because to their extensive technological competence. The entire software development process was planned as a shared learning opportunity between the startup team and freelancers.

Easier prototype solutions, like videos, made it simple to target customers. Due to close professional ties with freelancers, regional NGOs, and institutions, consumers were easily accessible. In some instances, university students run various student chapters (or student clubs) with the objective of bringing about social change. Startups were helped by these groups and were able to learn a lot about many problem areas. The rich support given by independent contractors enables them to quickly execute solutions and evolve them. Strategic collaborations boosted the level of trust, which eventually helped startups stay informed about new business prospects and improved their capacity to mobilize resources and continuously innovate. These dynamic innovation capabilities result in sustained business growth.

References

Gupta, V. (2021a). *Requirements engineering for social sector software applications.* Springer. https://doi.org/10.1007/978-3-030-83549-1

Gupta, V. (2021b). Requirement engineering challenges for social sector software development: Insights from multiple case studies. *Digital Government: Research and Practice, 2*(4), 1–13. https://doi.org/10.1145/3479982

Gupta, V., Fernández-Crehuet, J. M., & Gupta, C. (2022). Freelancer supported requirement engineering framework for software start-ups. *Computer, IEEE,.* https://doi.org/10.1109/MC.2022.3180711

Teaching Note

Synopsis

SocietyTech is an online startup that was established in 2016 with the goal of transforming society through social innovations. Startup (now a company) was jointly founded by Perter and James. The firm had to deal with a number of difficulties that were unique to the startup community, such as resource limitations and issues with social sectors, like user diversity, inadequate finance, and low revenues. In the last seven years, the firm has been able to both carry out social projects for other people and offer its technology solutions for social innovations. The business approach is built on enhancing customer value and ongoing cost savings. This example demonstrates how strategic collaborations with universities enabled it to cultivate relationships with clients and independent contractors for ongoing social advances.

Teaching Objectives

This case is suitable for startup management, innovation management, and strategy management courses at the undergraduate, graduate, and doctoral levels. Students will learn how to critically evaluate the effective open innovation approach of a real firm through this case, participate in insightful brainstorming sessions with their peers, and come up with improved strategies to innovate that could benefit the startup community. After the case study discussion is over, successful students will be able to:

- Critique startup innovation techniques for long-term competitive advantage.
- Critique the creations, developments, and executions of strategic partnerships.
- Integrate several components of the ecosystem of innovation for open innovation.

- Through a combination of outsourcing and internal development, critique the value proposition innovations.

Assignment Questions

This case will help entrepreneurs, students and researchers to replace their "hire as when required" approach with "forever together we can" approach.

Questions

Q1. To what extend the innovation strategy of Perter and James will work for long-term competitive advantage.

Q2. To what extend the academic partnerships will work for startup?

Q3. To what extend startup was able to integrate several components of the ecosystem of innovation for open innovation?

Q4. To what extend the startup was able to innovate value proposition through blend of outsourcing and in-house activities?

Teaching Plan Timing

The timing for a 120-min class is as follows (Table A.2).

Table A.2 Teaching plan timing

Topics	Timings (min)
Introduction	10
Analyzing the SOCIETYTECH context & challenges	20
Strategies adopted	45
Dynamic capabilities and business impacts	30
Key learning and postscript	15

Analysis of Assignment Questions

Since there is no one correct response, the one conceivable viewpoint is discussed in the paragraphs that follow.

Q1. **To what extend the innovation strategy of Peter and James will work for long-term competitive advantage.**

The tactic will undoubtedly result in a long-lasting competitive edge. Of course, the market performance and financial indicators provided in the instance make this obvious. However, the startup is able to get over its incapacity to access the knowledge sources thanks to strategic backing from freelancers, clients, and academia. Better support for new idea invention, execution, and commercialization is the overall effect. With the support of these collaborations, the startup has improved its reputation and connections with freelancers, clients, and academic institutions, enabling it to make the greatest use of their tactical knowledge and transform them into its strategic resources. Due of the strategic resources it currently possesses, the rival startup will struggle to replace it.

Q2. **To what extend the academic partnerships will work for startup?**

This is demonstrated by the fact that the startup first collaborated with colleges before beginning to work with clients and independent contractors. It made use of the universities' significant professional and social reach. The goal of fostering entrepreneurship is strongly aligned with the startup business strategy and the university's "third mission." These alliances are more focused on achieving corporate goals, like the third mission, than they are on making money. Since the revenues and earnings in the social sector are so low, this is implied. Startups can obtain all the resources they need from academia, such as student interns, research, academic specialists, involvement in social campaigns, branding, and access to its professional networks with financing organizations, government agencies, and institutions from the social sector.

Q3. **To what extend startup was able to integrate several components of the ecosystem of innovation for open innovation?**

A startup was able to bring together customers, academics, and freelancers around the unifying theme of making a difference in society. With academic assistance, it acquired the tools it needed to inspire clients and freelancers to support it. With help from the university, it was able to grow its professional connections with organizations like NGOs, which made it easier for it to connect with new clients. The capacity to thoroughly investigate the problem domain and produce a proven solution was the result. Additionally, it was possible to boost the self-assurance of independent contractors by assisting them in connecting their career, for instance through industrial doctorates, with startup. In this manner, it was able to combine them with a more strategic vision and social results.

Q4. **To what extend the startup was able to innovate value proposition through blend of outsourcing and in-house activities?**

Startup was able to reduce its reliance on complete outsourcing. The emphasis was on collaborative learning and working. By working with freelancers, it was able to build the skills of its startup team and prevent the loss of any tactical expertise that might have been held by freelancers. Together, the startup team was able to apply the knowledge gained from earlier projects to current ones, which are frequently incremental and iterative, i.e. constantly evolving value propositions.

Key Learnings

The instructor could summarize what the SocietyTech case reveals in order to wrap up the case discussion.

- Strategic collaborations with components of the innovation ecosystem are crucial for developing flexible capacities for ongoing innovation.
- The value of investing resources in accordance with a solid innovation strategy.
- Blended outsourced and internal activities are crucial for collaborative learning.
- Importance of converting resources into strategic resources.

Postscript

Although the startup is present in several nations, it is currently concentrating on more aggressive global market expansion. Priority is given to collaborating with international universities before moving on to their current university partners. Through relationships with overseas universities, it has access to foreign institutions for market research and to local organizations like NGOs. A good technique to get around the so-called "liability of foreignness" is to greatly simplify consumer access. Through its collaborations with overseas colleges, it is expanding its freelancing platforms and offering additional career support to students. In order to give its freelancers access to more specialized abilities, the business will also offer licenses for online learning environments like Coursera and Edx. In order to promote long-term relationships, additional professional opportunities are also being identified. According to the founder of the startup, the pandemic presented chances for the business in the form of growing trends in the gig economy, with more freelancers readily available across a variety of freelancing platforms. They are now starting to find qualified freelancers on various platforms, delegate a few jobs, and then incorporate them into the startup company's platform.

Printed in the United States
by Baker & Taylor Publisher Services